CHILDREN ON THE FARM

A Postcard Book of Photographs by Pete Wettach

Selected by Leslie A. Loveless

PUBLISHED BY THE

UNIVERSITY OF IOWA PRESS

FOR THE STATE HISTORICAL SOCIETY OF IOWA

A Bur Oak Book

University of Iowa Press, Iowa City 52242

Design by Karen Copp

http://www.uiowapress.org

International Standard Book Number 0-87745-913-4

Printed in China

10 9 8 7 6 5 4 3 2 1

Printed on acid-free paper

During the Great Depression, World War II, and the postwar years, no place was more emblematic of home than the American family farm.

Iowa photographer A. M. "Pete" Wettach (1901–1976) knew and loved farm life. He traveled the countryside administering a Farm Security Administration relief program but also taking photographs of people he met along the way. Many of his most striking photographs include children—at play, at work, doing their chores, or exploring and learning.

I feel as if I know Pete Wettach well, although he died more than a quarter of a century ago. My chance discovery of a few of his photos led me to a massive collection of images of midwestern farm life, some of which I put together for the book *A Bountiful Harvest*. This collection of postcards includes my favorite photographs from that book as well as some never before published. I hope you enjoy them as much as I do.

Leslie A. Loveless

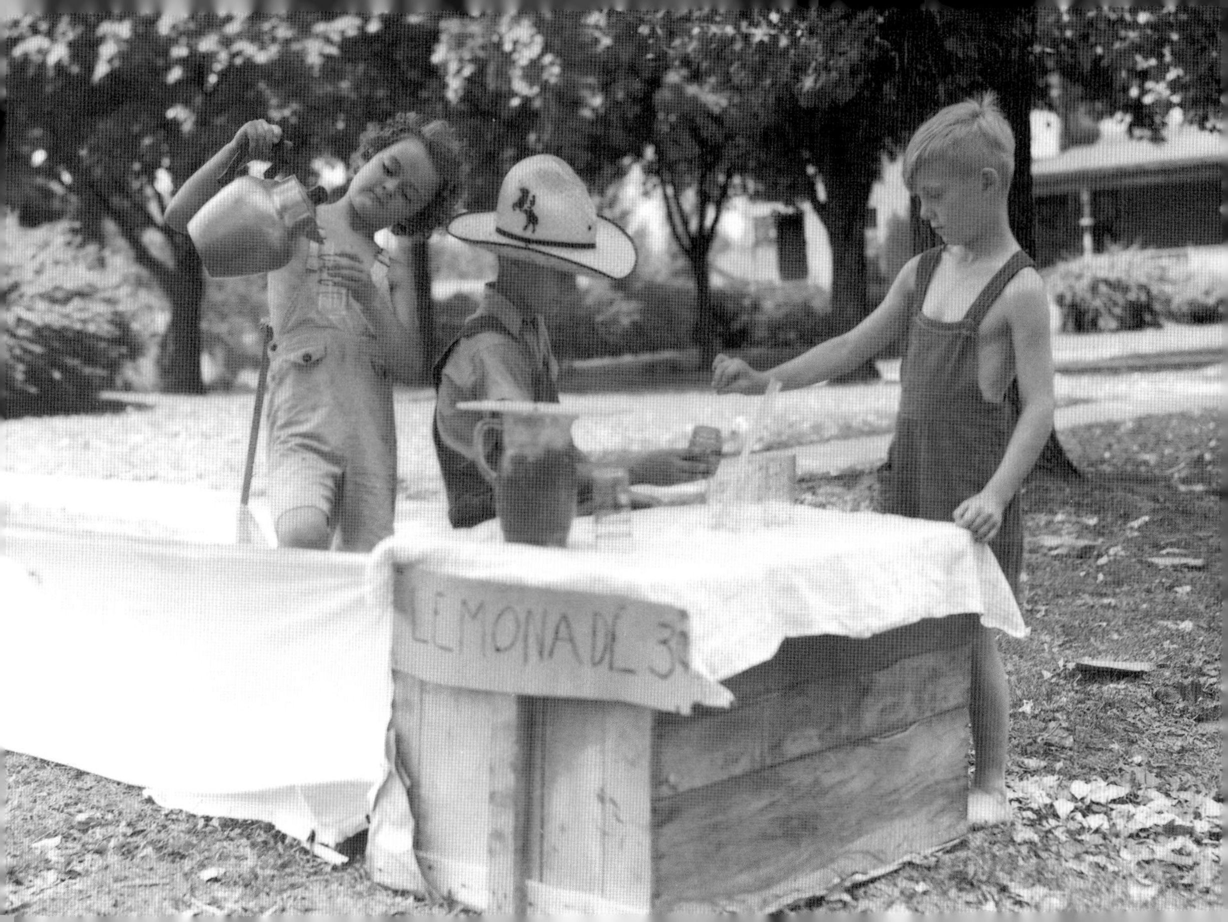
LEMONADE 3¢

CHILDREN ON THE FARM

Selling lemonade, Mt. Pleasant, Iowa, 1930s. Photo by A. M. "Pete" Wettach. State Historical Society of Iowa, Iowa City.

USE

FIRST-CLASS

POSTAGE

University of Iowa Press http://www.uiowapress.org

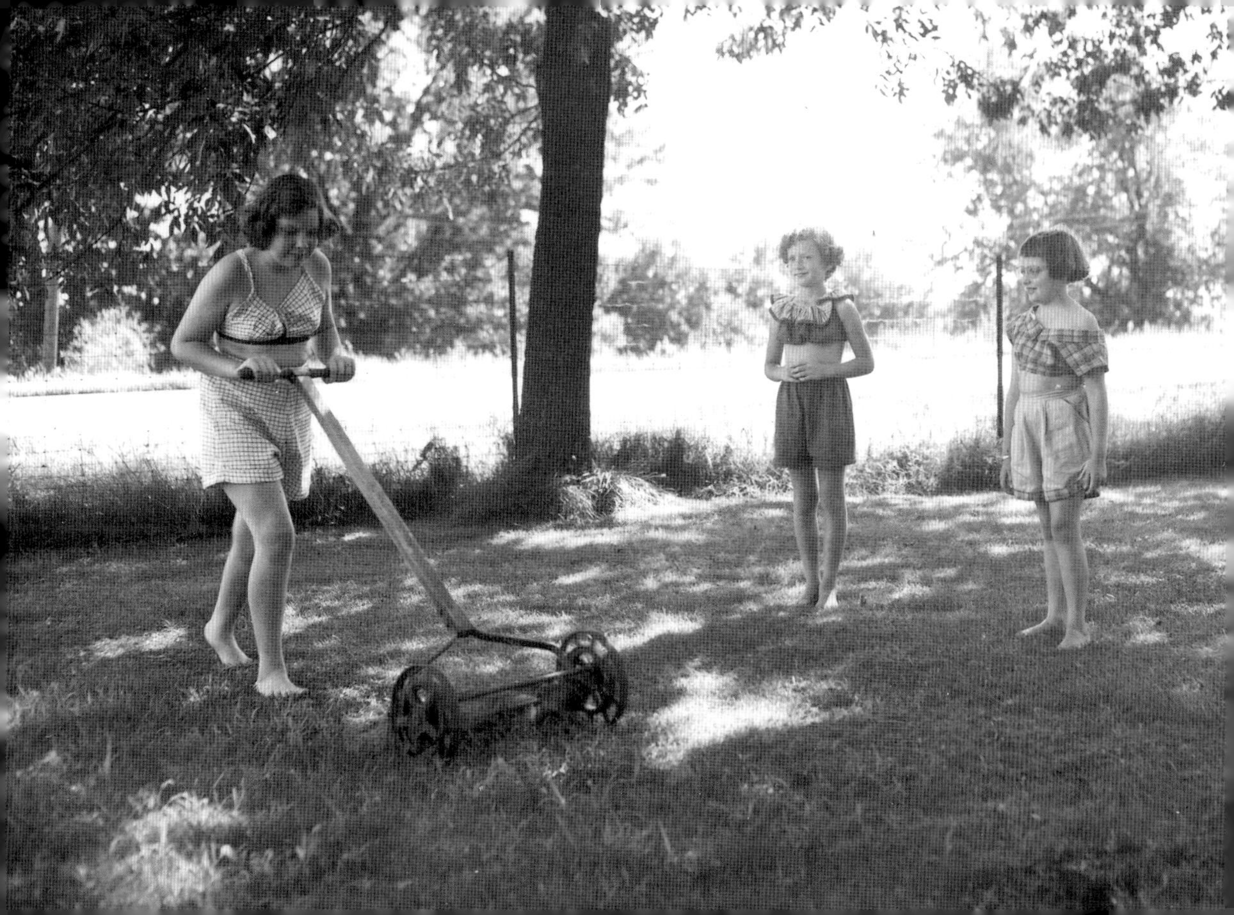

CHILDREN ON THE FARM

Mowing the lawn. Photo by A. M. "Pete" Wettach.
State Historical Society of Iowa, Iowa City.

USE

FIRST-CLASS

POSTAGE

University of Iowa Press http://www.uiowapress.org

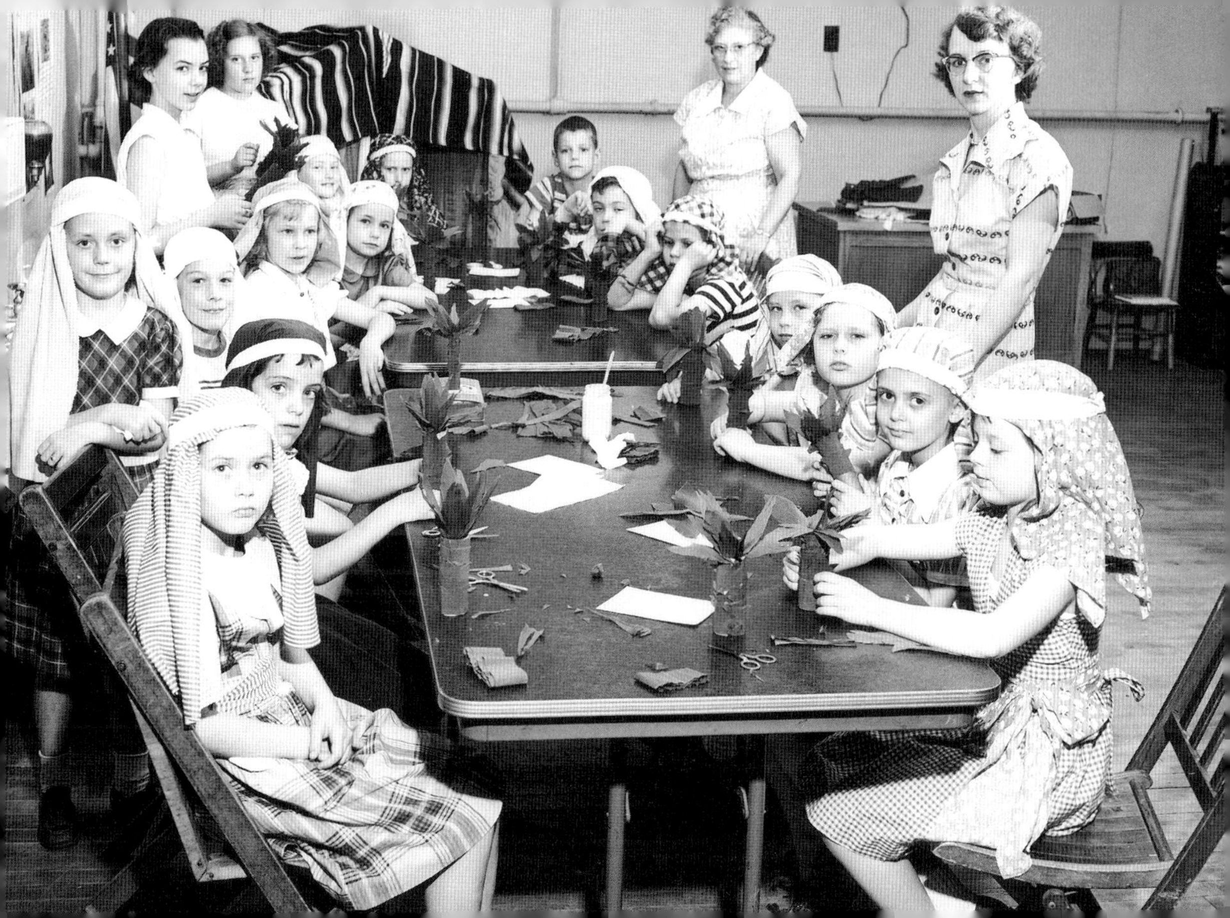

CHILDREN ON THE FARM

Sunday school students and teachers. Photo by A. M. "Pete" Wettach. State Historical Society of Iowa, Iowa City.

USE

FIRST-CLASS

POSTAGE

University of Iowa Press http://www.uiowapress.org

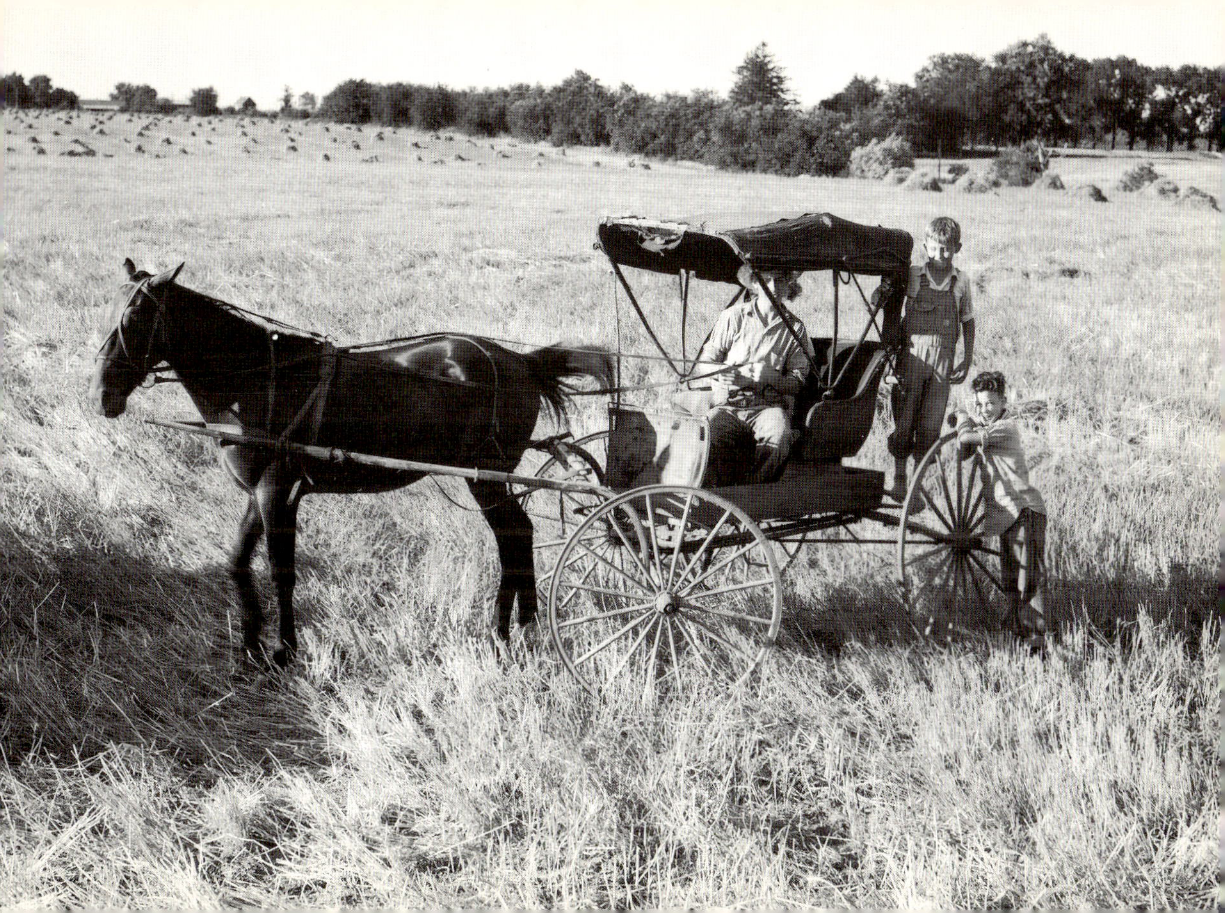

CHILDREN ON THE FARM

Out for a spin in an antique buggy, 1930s. Photo by A. M. "Pete" Wettach. State Historical Society of Iowa, Iowa City.

USE

FIRST-CLASS

POSTAGE

University of Iowa Press http://www.uiowapress.org

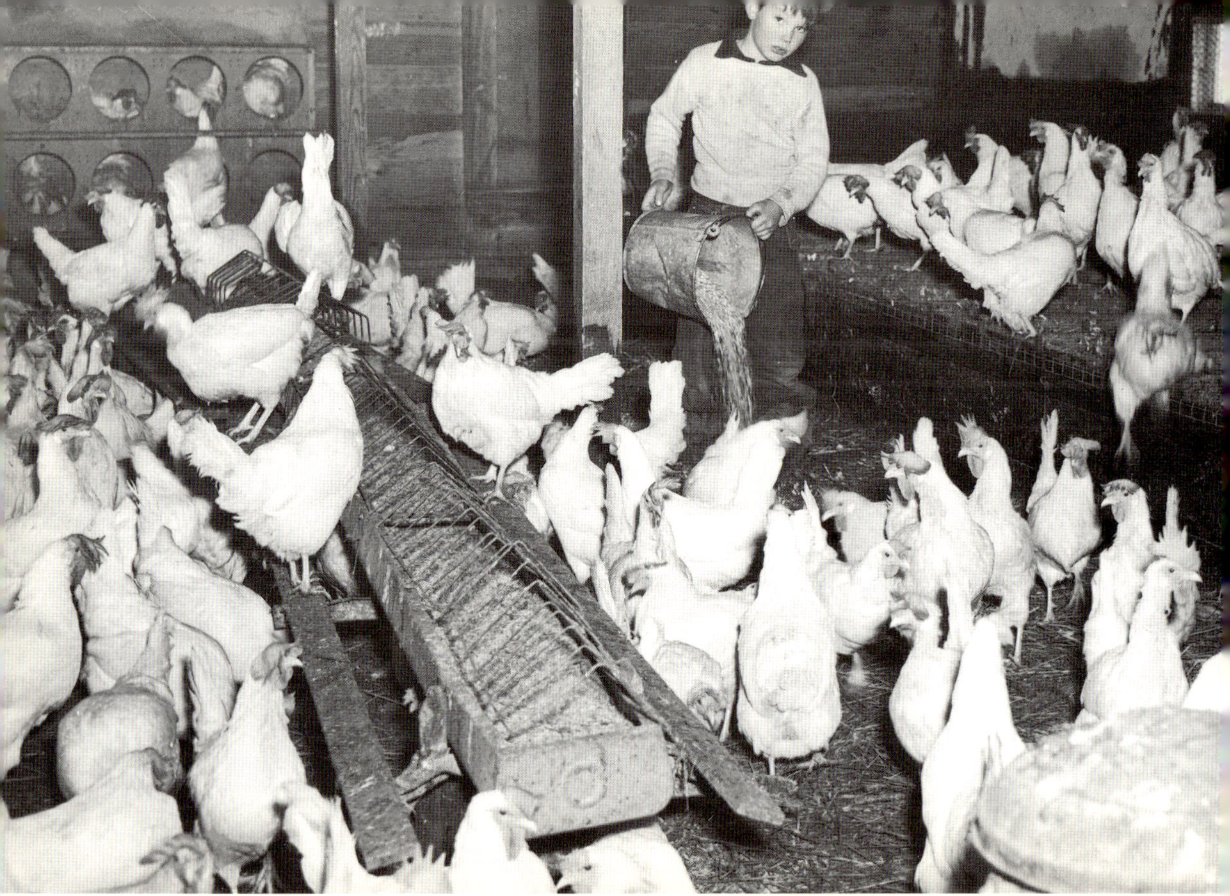

CHILDREN ON THE FARM

Feeding the chickens. Photo by A. M. "Pete" Wettach.
State Historical Society of Iowa, Iowa City.
Photo courtesy Reiman Publications.

USE
FIRST-CLASS
POSTAGE

University of Iowa Press http://www.uiowapress.org

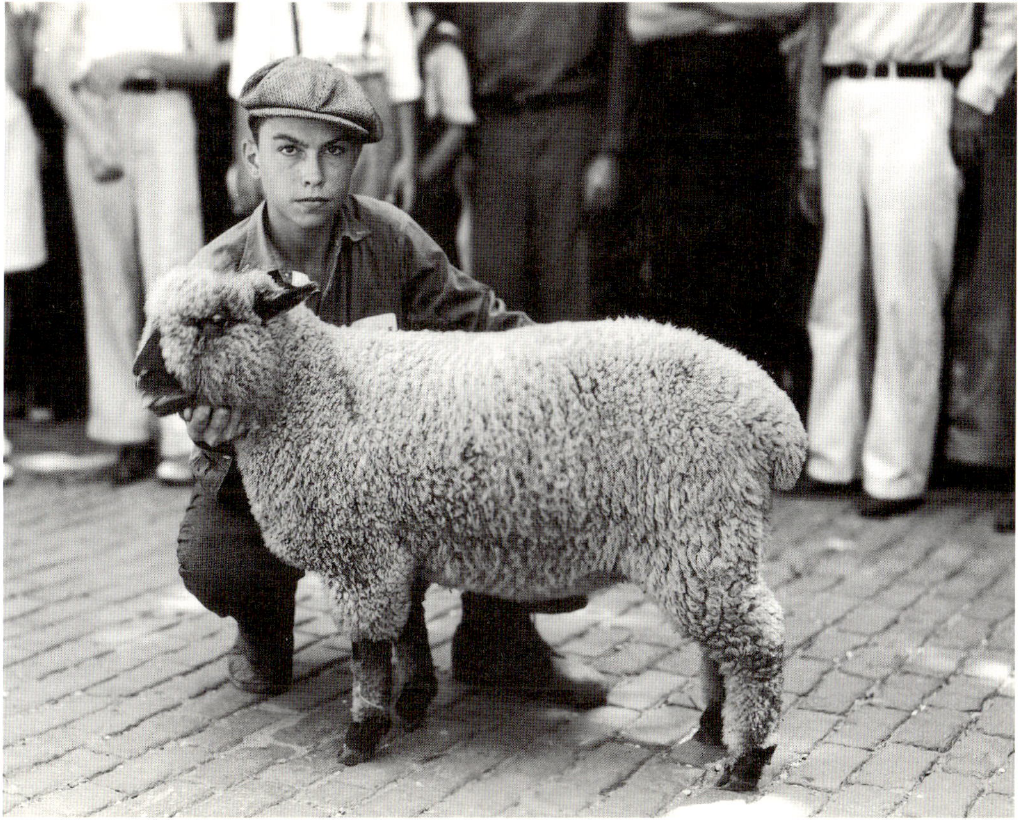

CHILDREN ON THE FARM

Grand Champion lamb, 4-H Lamb Day, Henry County, Iowa. Photo by A. M. "Pete" Wettach. State Historical Society of Iowa, Iowa City.

USE

FIRST-CLASS

POSTAGE

University of Iowa Press http://www.uiowapress.org

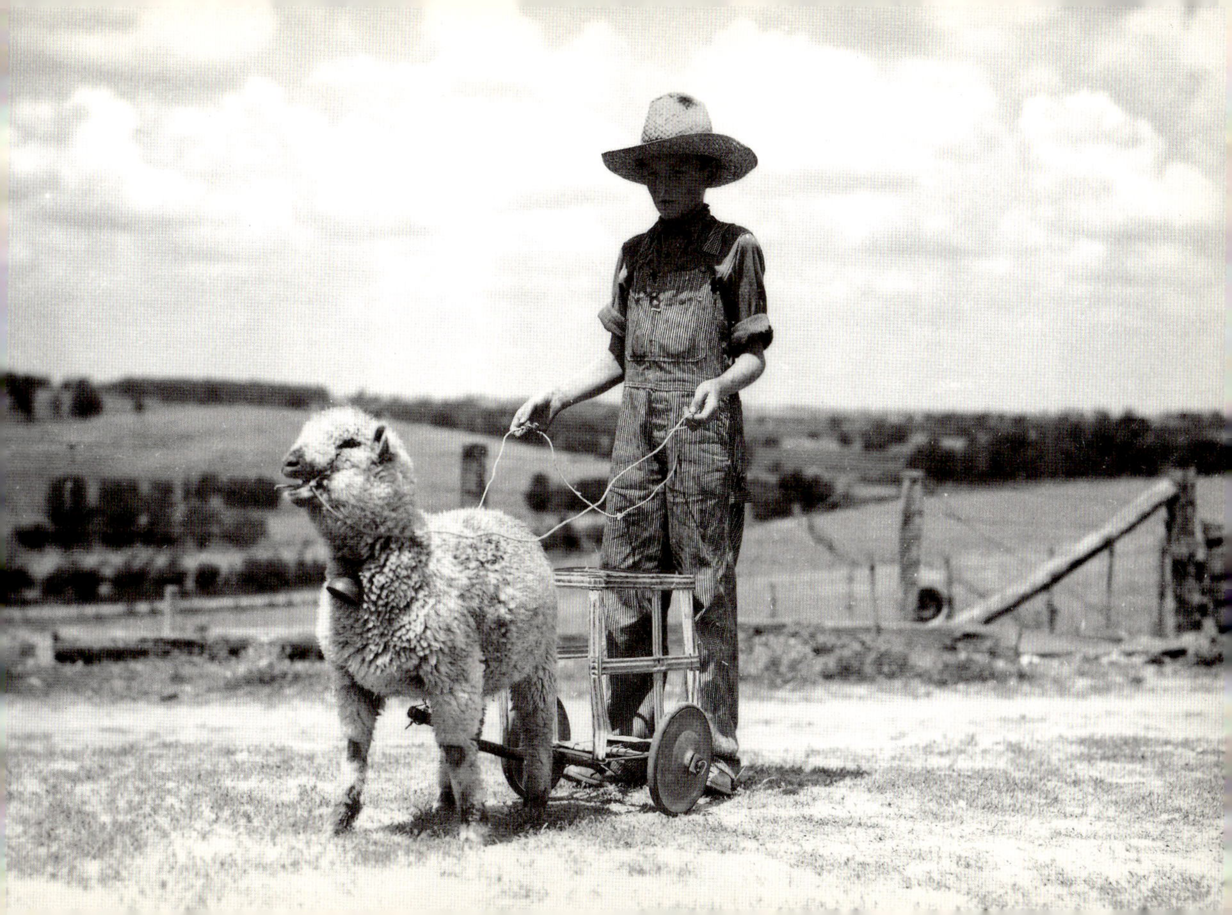

CHILDREN ON THE FARM

"Off for the races," Davis County, Iowa. Photo by A. M. "Pete" Wettach. State Historical Society of Iowa, Iowa City.

USE

FIRST-CLASS

POSTAGE

University of Iowa Press http://www.uiowapress.org

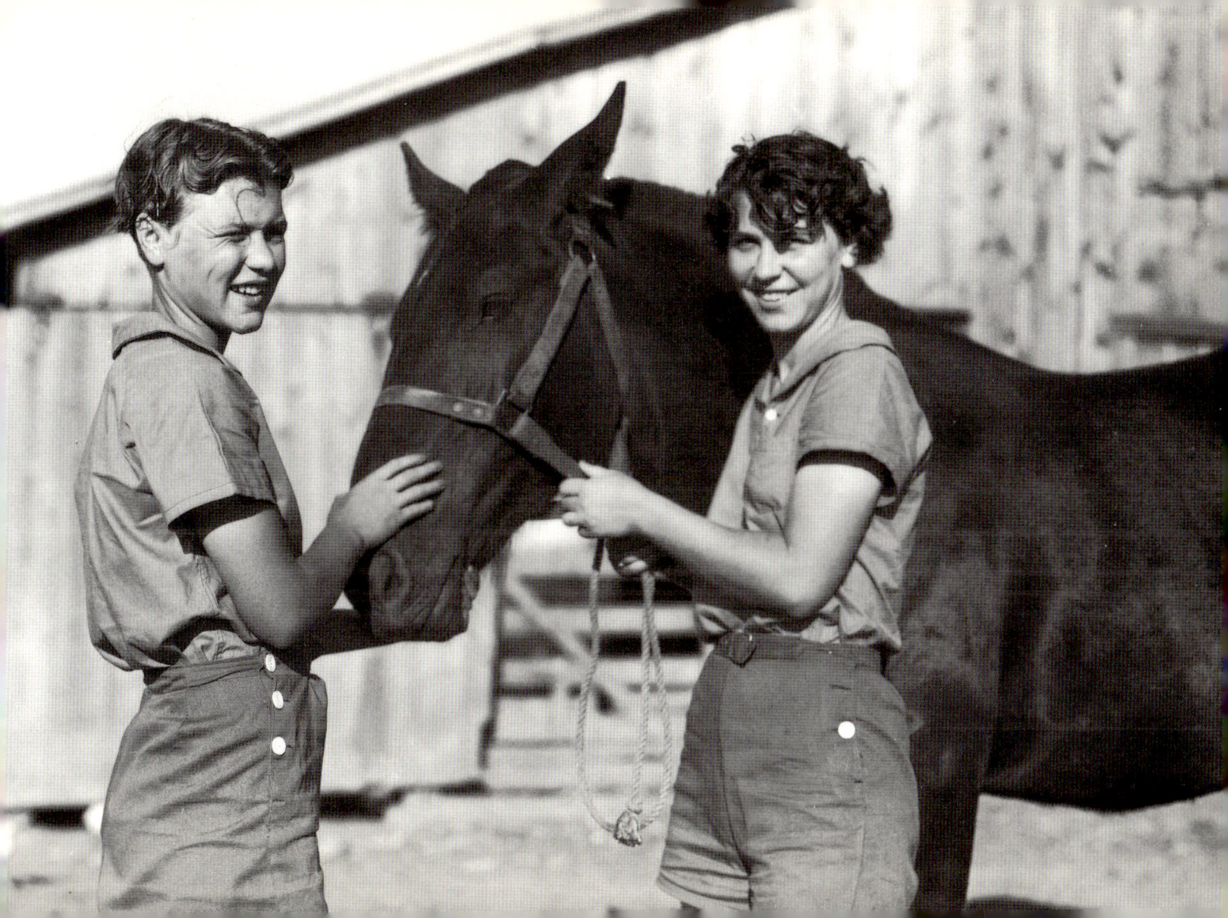

CHILDREN ON THE FARM

Two sisters, Henry County, Iowa, 1939. Photo by A. M. "Pete" Wettach. State Historical Society of Iowa, Iowa City.

USE

FIRST-CLASS

POSTAGE

University of Iowa Press http://www.uiowapress.org

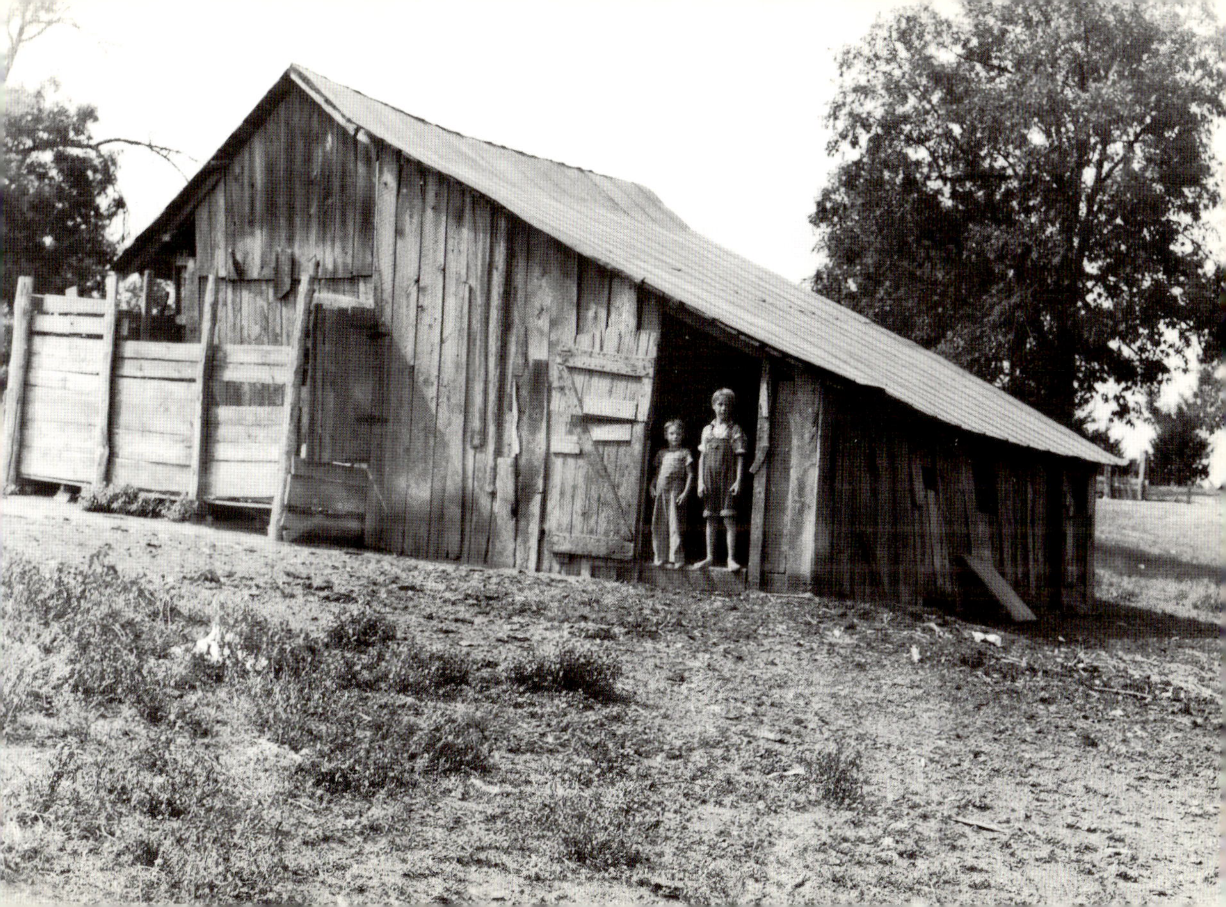

CHILDREN ON THE FARM

Two children in the doorway of an old barn. Photo by A. M. "Pete" Wettach. State Historical Society of Iowa, Iowa City.

USE

FIRST-CLASS

POSTAGE

University of Iowa Press http://www.uiowapress.org

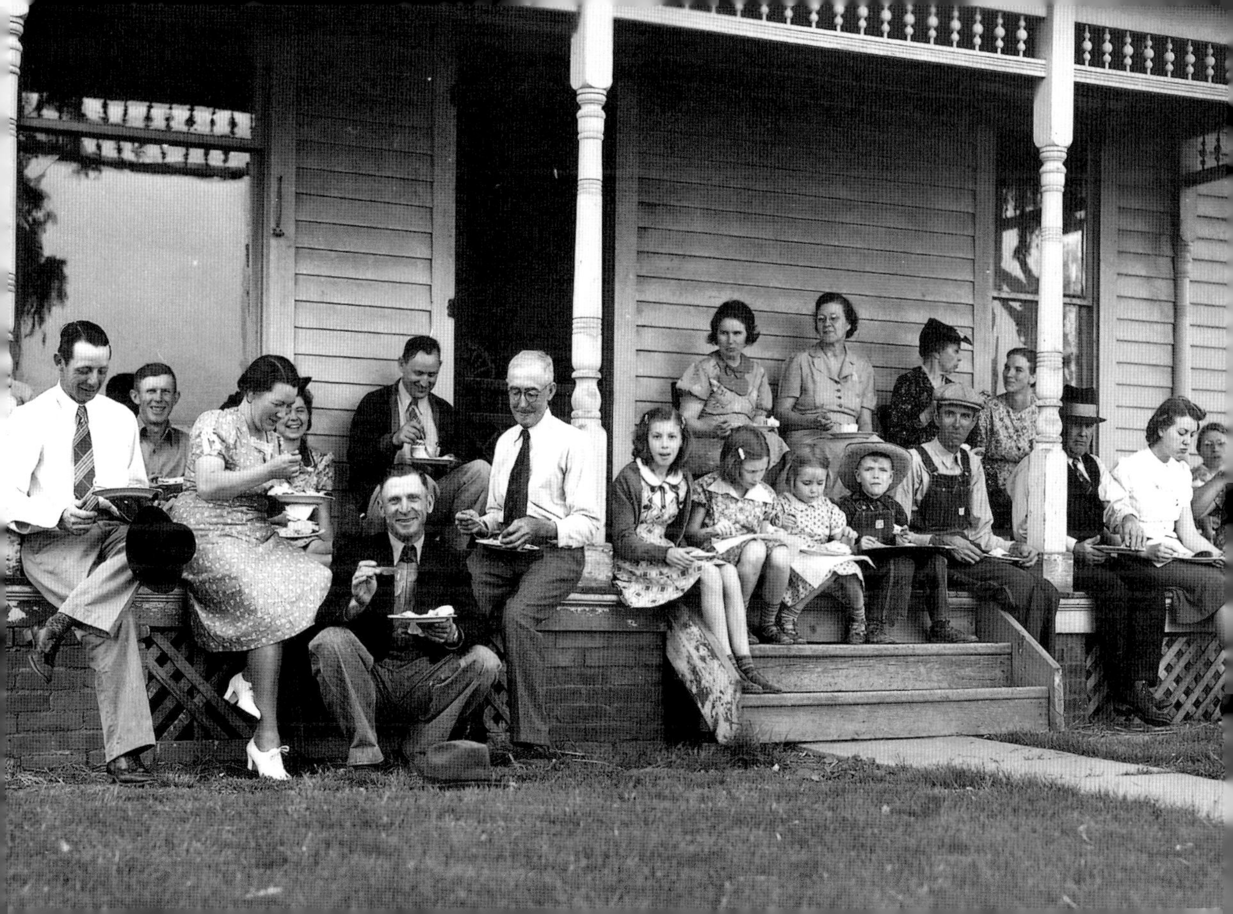

CHILDREN ON THE FARM

Picnic for Farm Security Administration client families from Jefferson County, Iowa, ca. 1941. Photo by A. M. "Pete" Wettach. State Historical Society of Iowa, Iowa City.

USE

FIRST-CLASS

POSTAGE

University of Iowa Press http://www.uiowapress.org

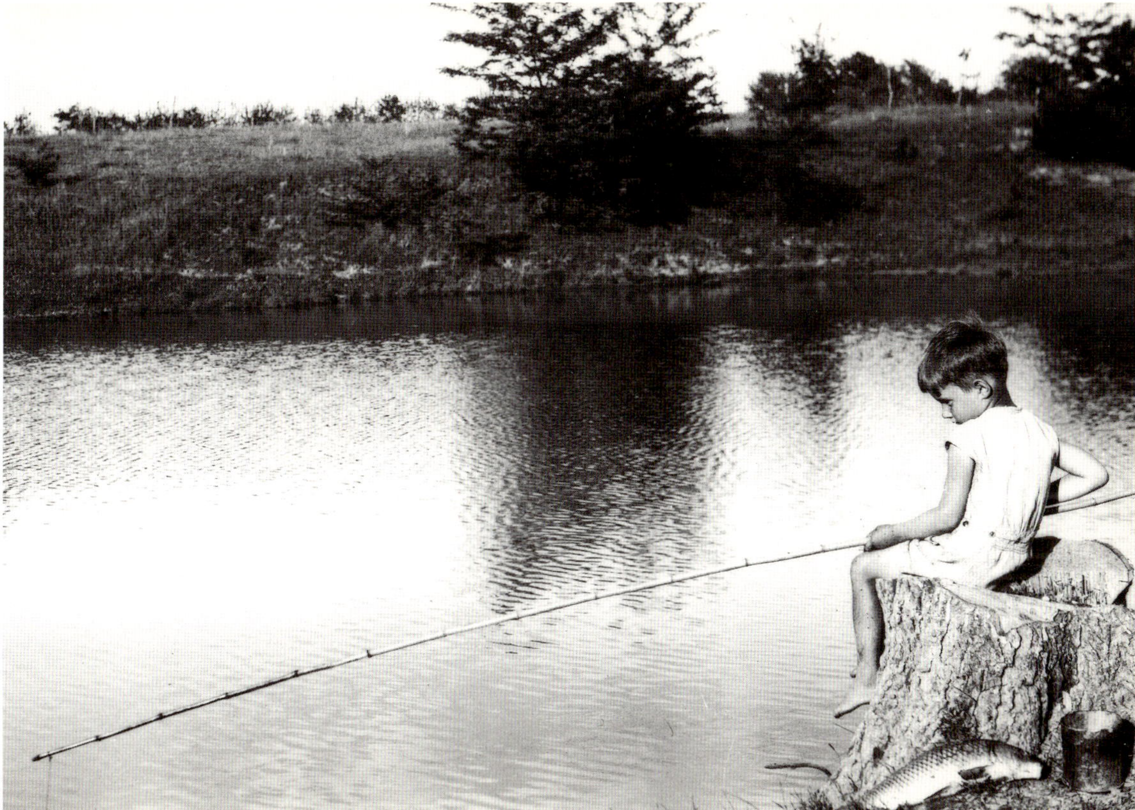

CHILDREN ON THE FARM

Fishing, 1930s. Photo by A. M. "Pete" Wettach.
State Historical Society of Iowa, Iowa City.

USE

FIRST-CLASS

POSTAGE

University of Iowa Press http://www.uiowapress.org

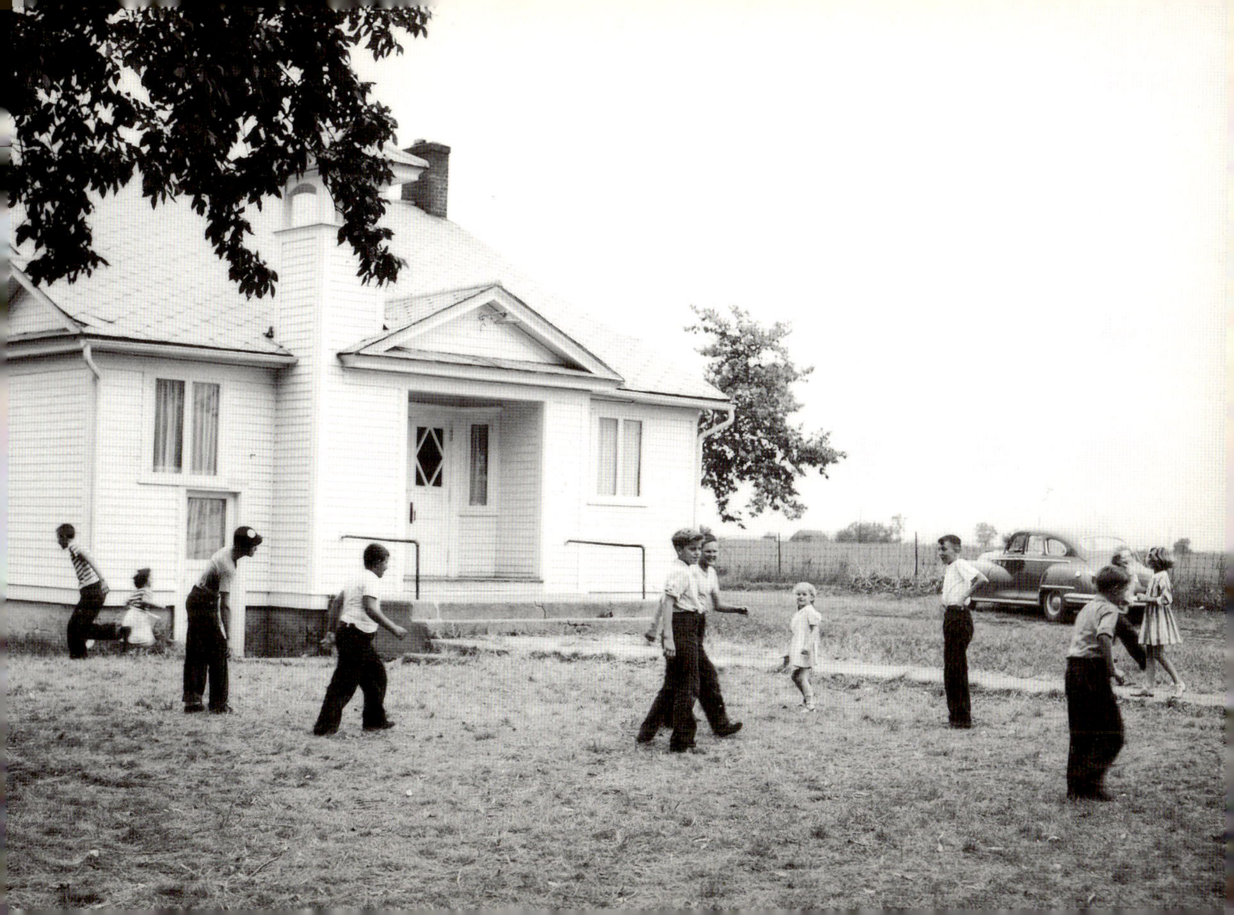

CHILDREN ON THE FARM

Playing in the schoolyard. Photo by A. M. "Pete" Wettach.
State Historical Society of Iowa, Iowa City.

USE

FIRST-CLASS

POSTAGE

University of Iowa Press http://www.uiowapress.org

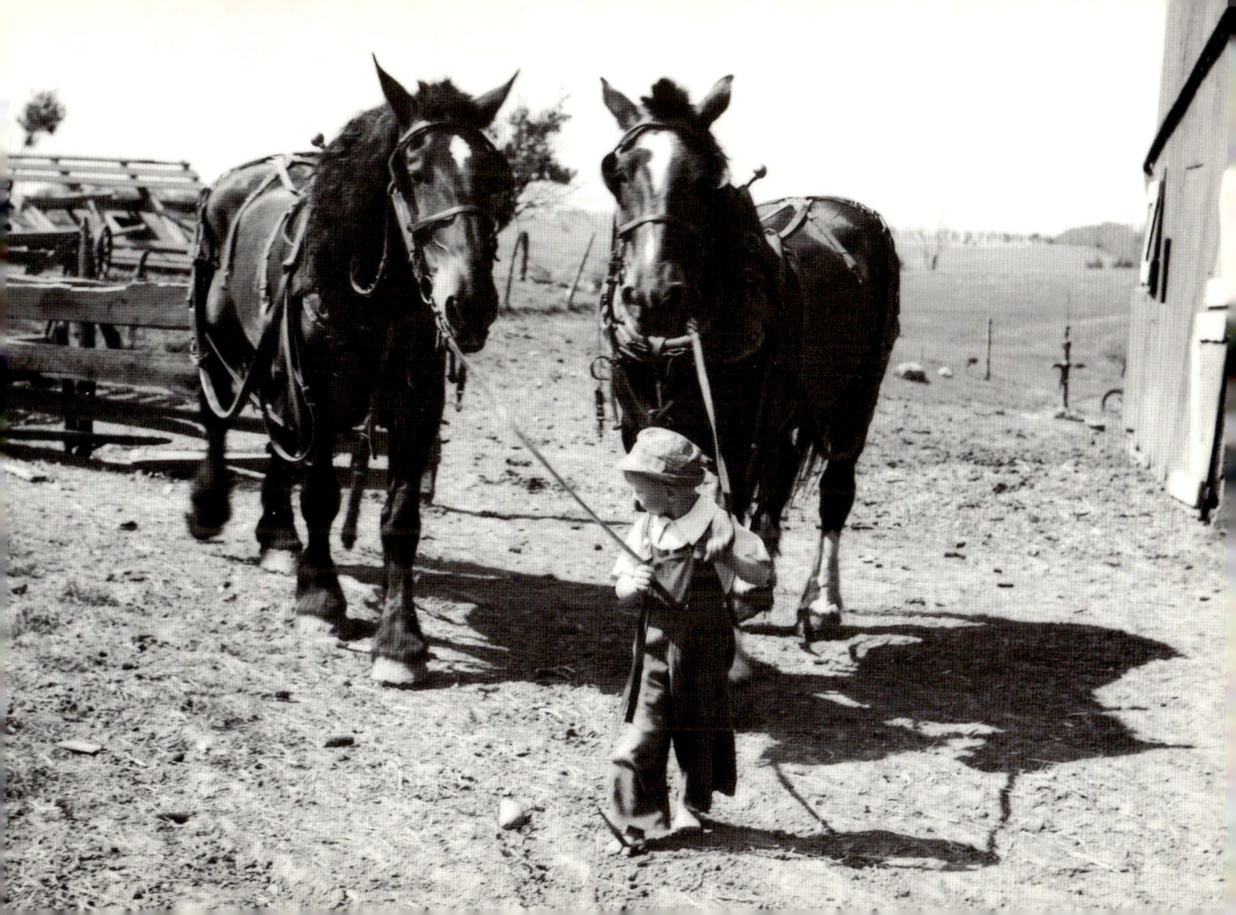

CHILDREN ON THE FARM

Leading two draft horses, 1930s. Photo by A. M. "Pete" Wettach. State Historical Society of Iowa, Iowa City.

USE

FIRST-CLASS

POSTAGE

University of Iowa Press   http://www.uiowapress.org

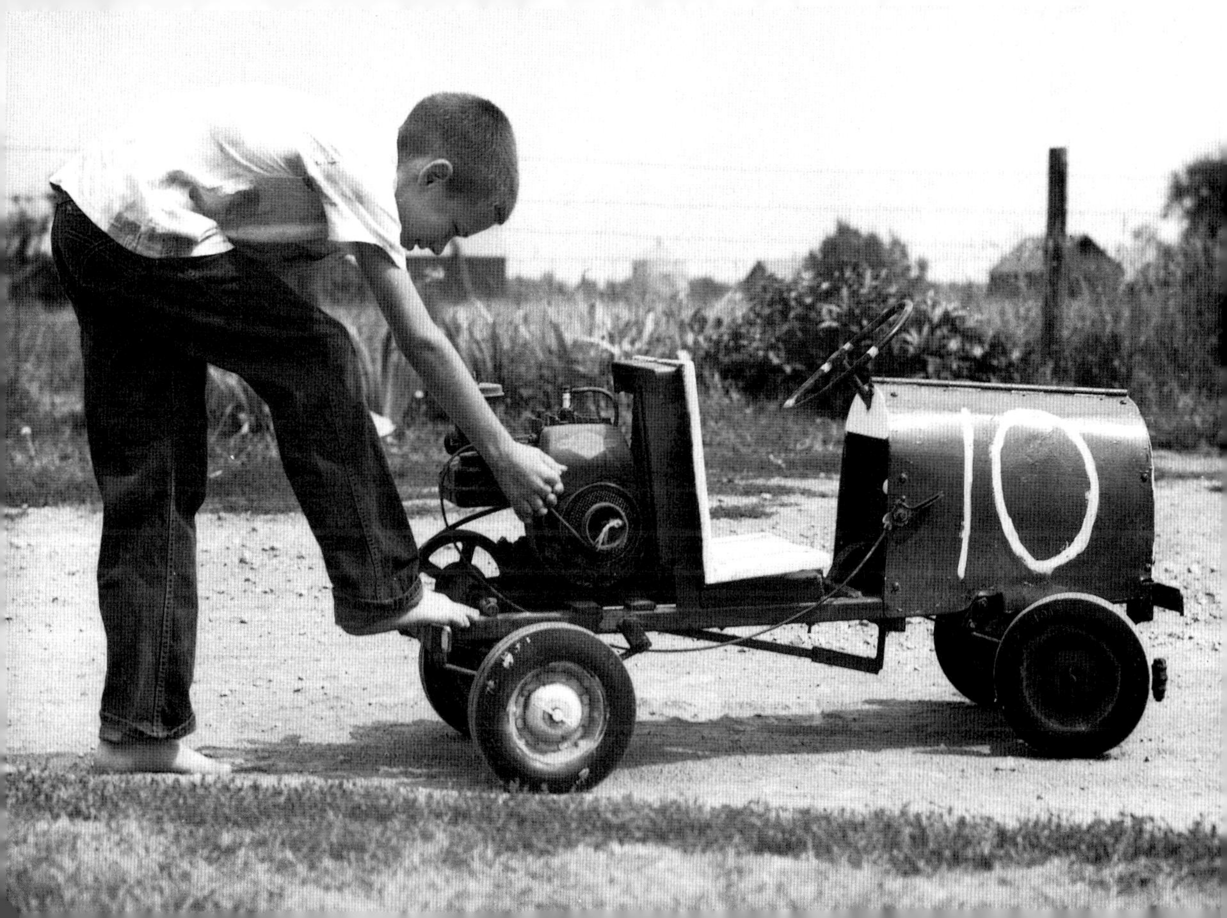
10

CHILDREN ON THE FARM

A boy starts a racer built by his dad, mid 1950s. Photo by A. M. "Pete" Wettach. State Historical Society of Iowa, Iowa City.

USE

FIRST-CLASS

POSTAGE

University of Iowa Press http://www.uiowapress.org

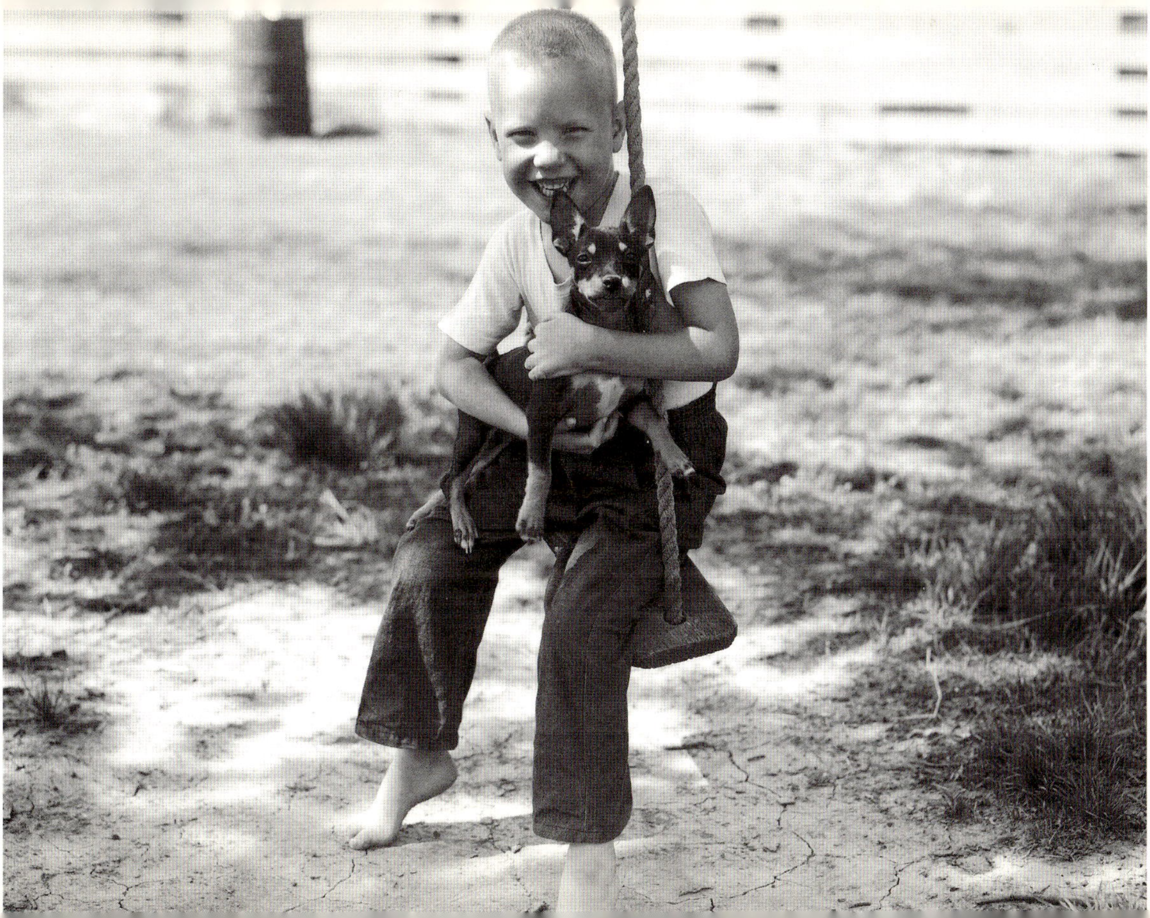

CHILDREN ON THE FARM

A boy and his dog, 1950s. Photo by A. M. "Pete" Wettach.
State Historical Society of Iowa, Iowa City.

USE

FIRST-CLASS

POSTAGE

University of Iowa Press   http://www.uiowapress.org

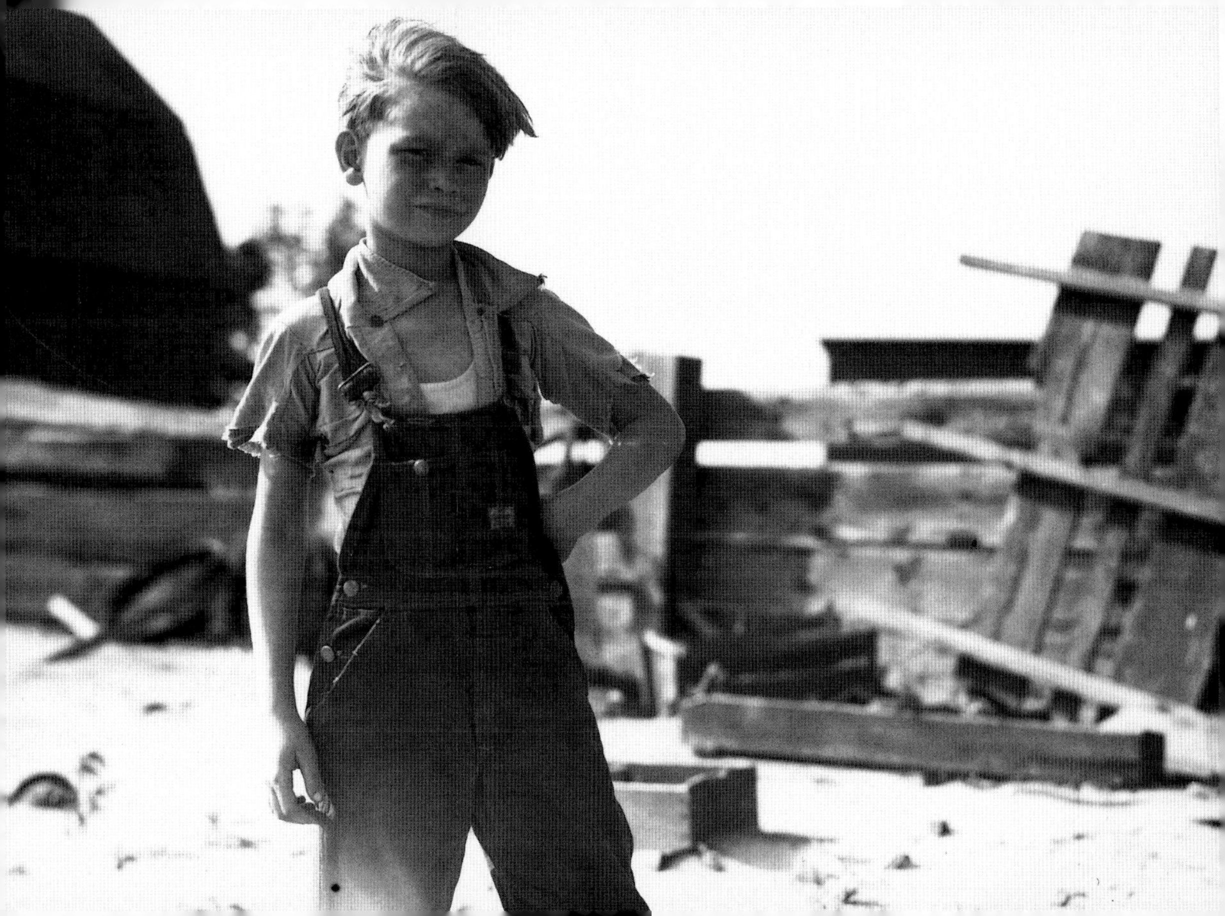

CHILDREN ON THE FARM

Standing in the barnyard. Photo by A. M. "Pete" Wettach.
State Historical Society of Iowa, Iowa City.

USE

FIRST-CLASS

POSTAGE

University of Iowa Press http://www.uiowapress.org

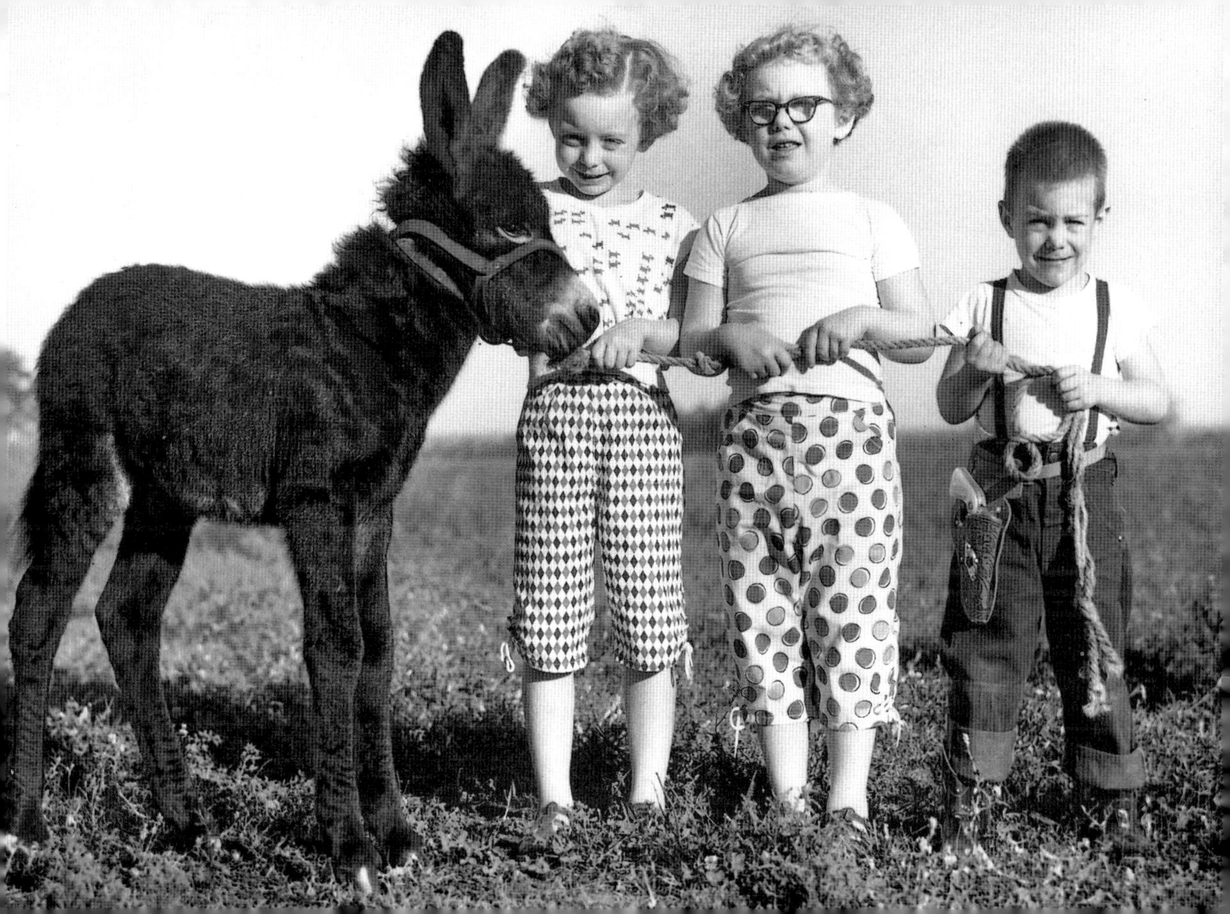

CHILDREN ON THE FARM

Three children and their pet burro colt, 1956. Photo by A. M. "Pete" Wettach. State Historical Society of Iowa, Iowa City.

USE

FIRST-CLASS

POSTAGE

University of Iowa Press http://www.uiowapress.org

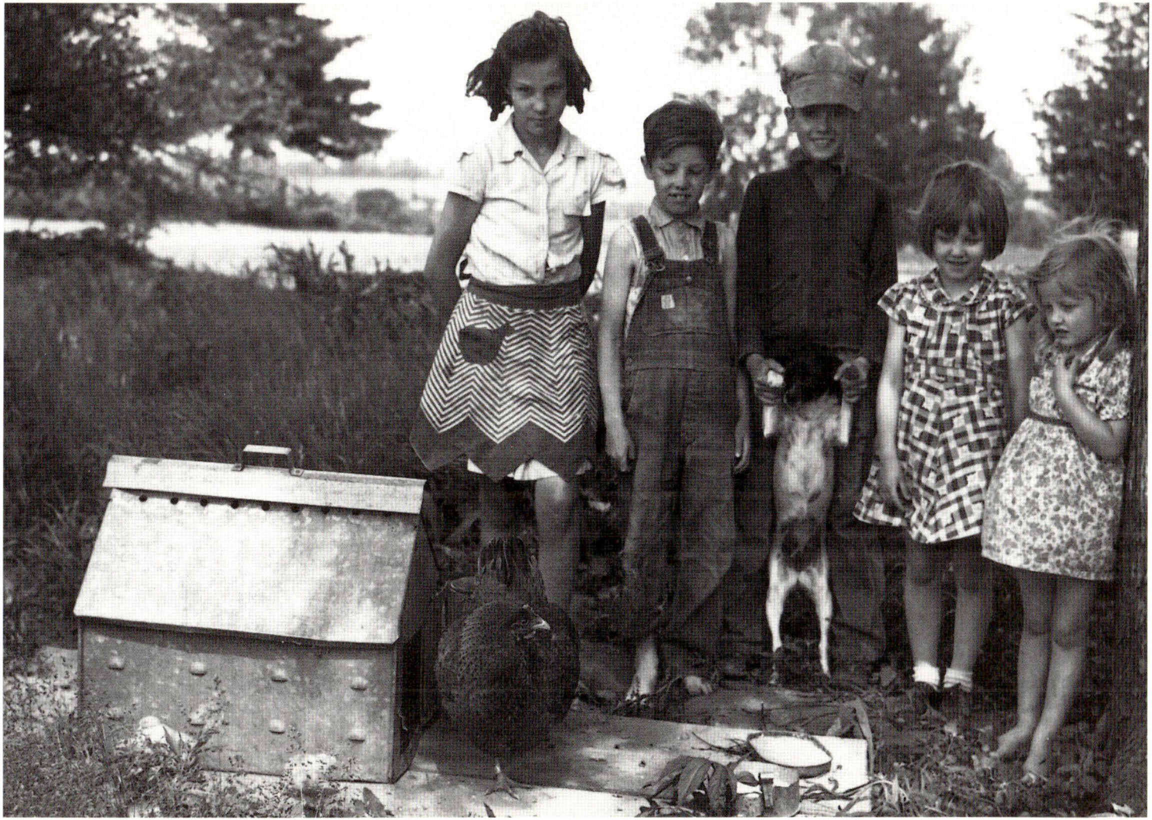

CHILDREN ON THE FARM

Five siblings and their pet hen, 1940s. Photo by A. M. "Pete" Wettach. State Historical Society of Iowa, Iowa City.

USE

FIRST-CLASS

POSTAGE

University of Iowa Press http://www.uiowapress.org

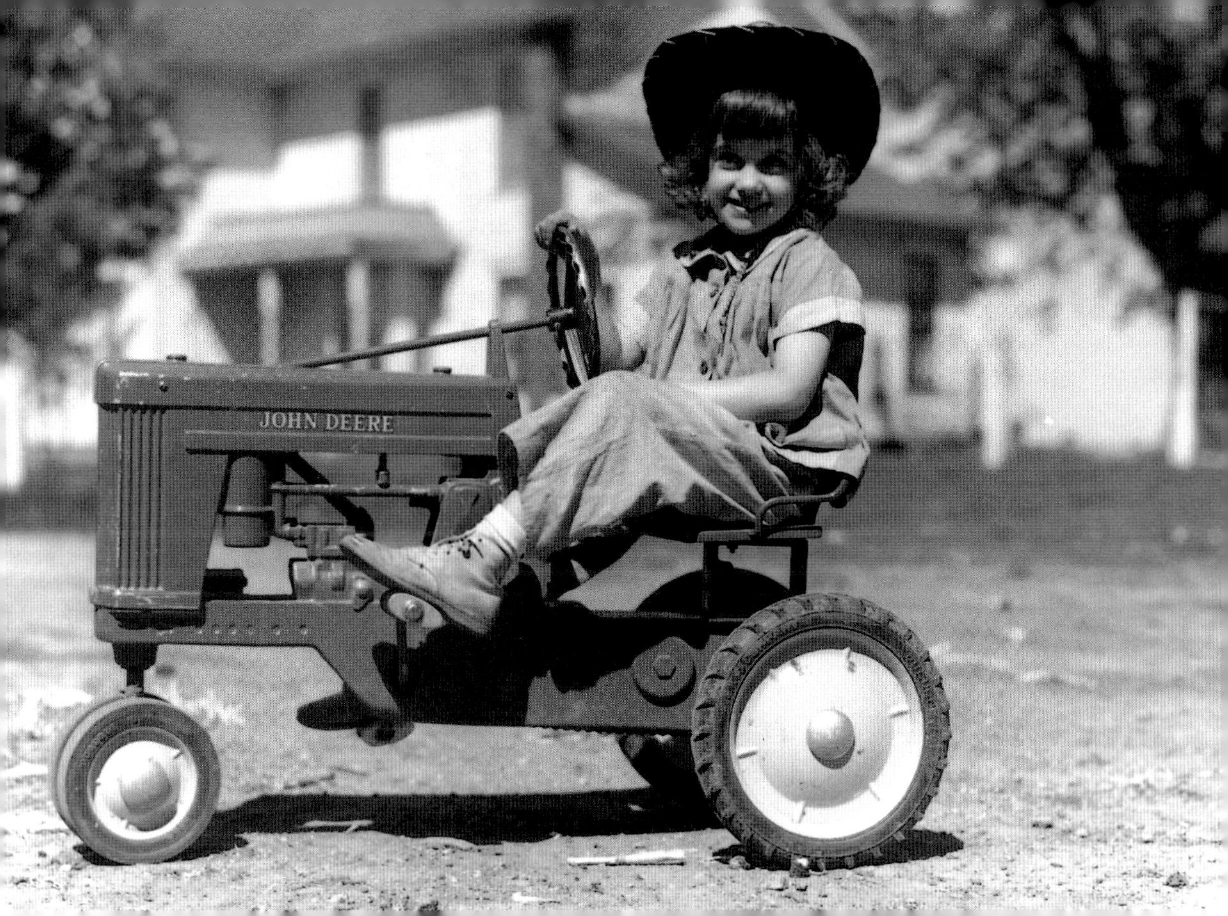
JOHN DEERE

CHILDREN ON THE FARM

A girl on her pedal tractor, 1955. Photo by A. M. "Pete" Wettach. State Historical Society of Iowa, Iowa City.

USE

FIRST-CLASS

POSTAGE

University of Iowa Press http://www.uiowapress.org

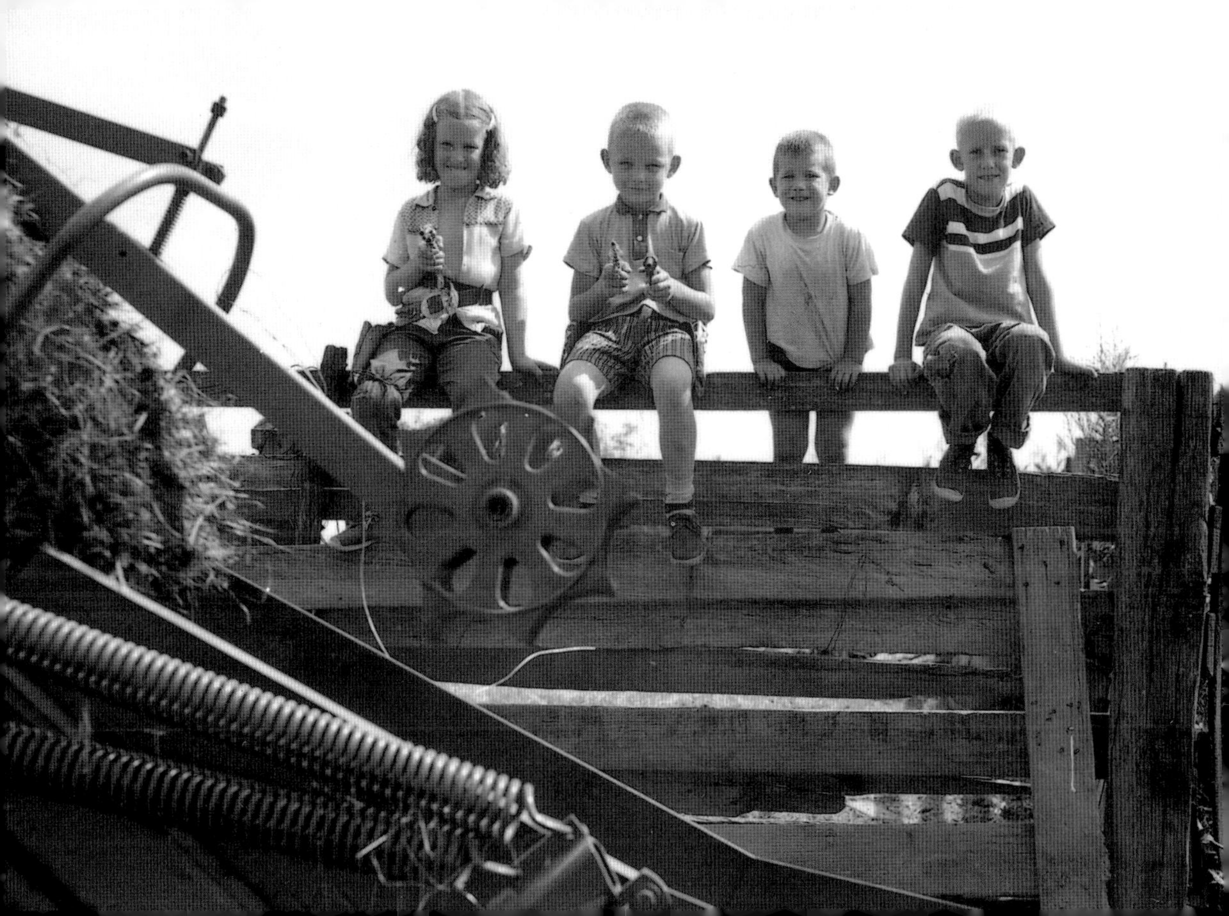

CHILDREN ON THE FARM

Farm kids play with their city cousins, ca. 1960. Photo by A. M. "Pete" Wettach. State Historical Society of Iowa, Iowa City.

USE

FIRST-CLASS

POSTAGE

University of Iowa Press http://www.uiowapress.org

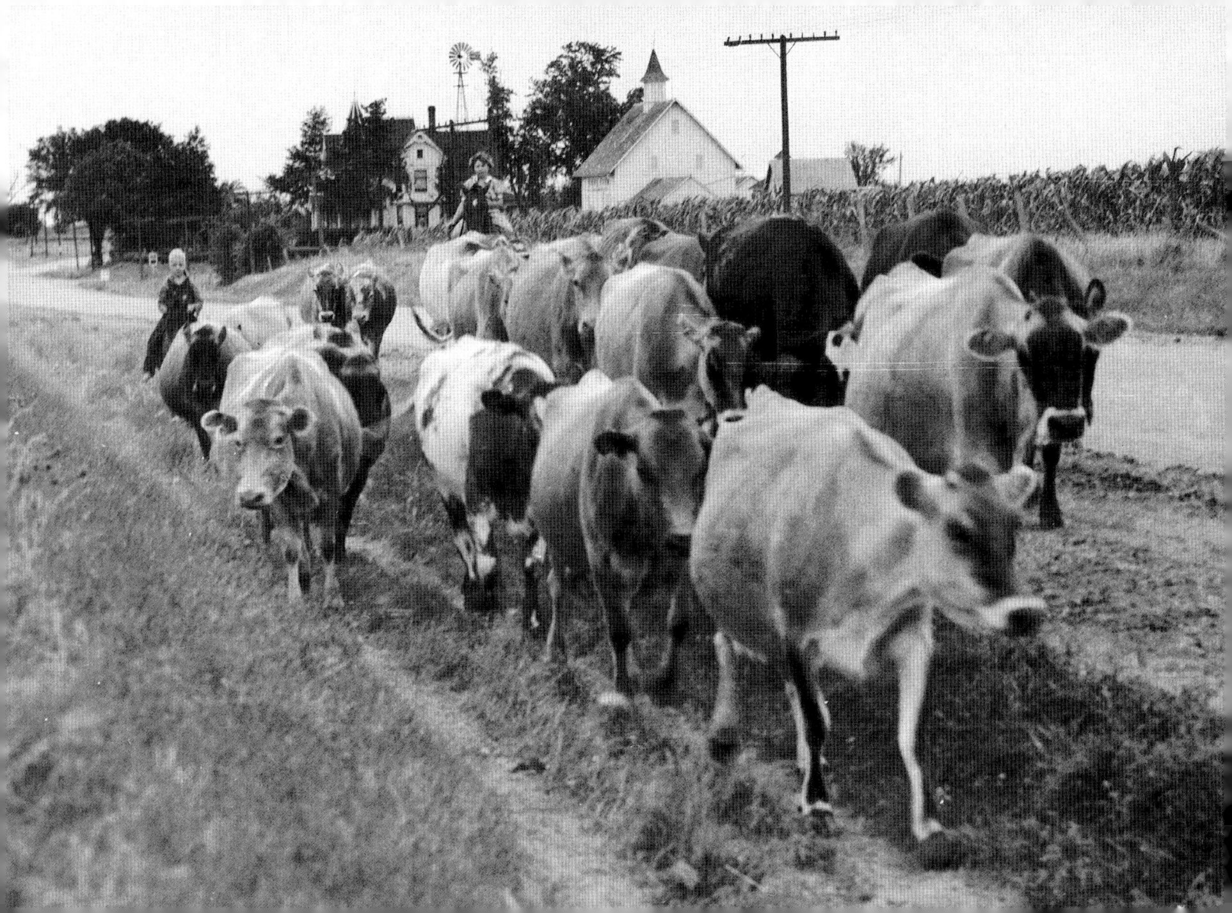

CHILDREN ON THE FARM

Herding cattle down a country road. Photo by A. M. "Pete" Wettach. State Historical Society of Iowa, Iowa City. Photo courtesy Reiman Publications.

USE

FIRST-CLASS

POSTAGE

University of Iowa Press http://www.uiowapress.org

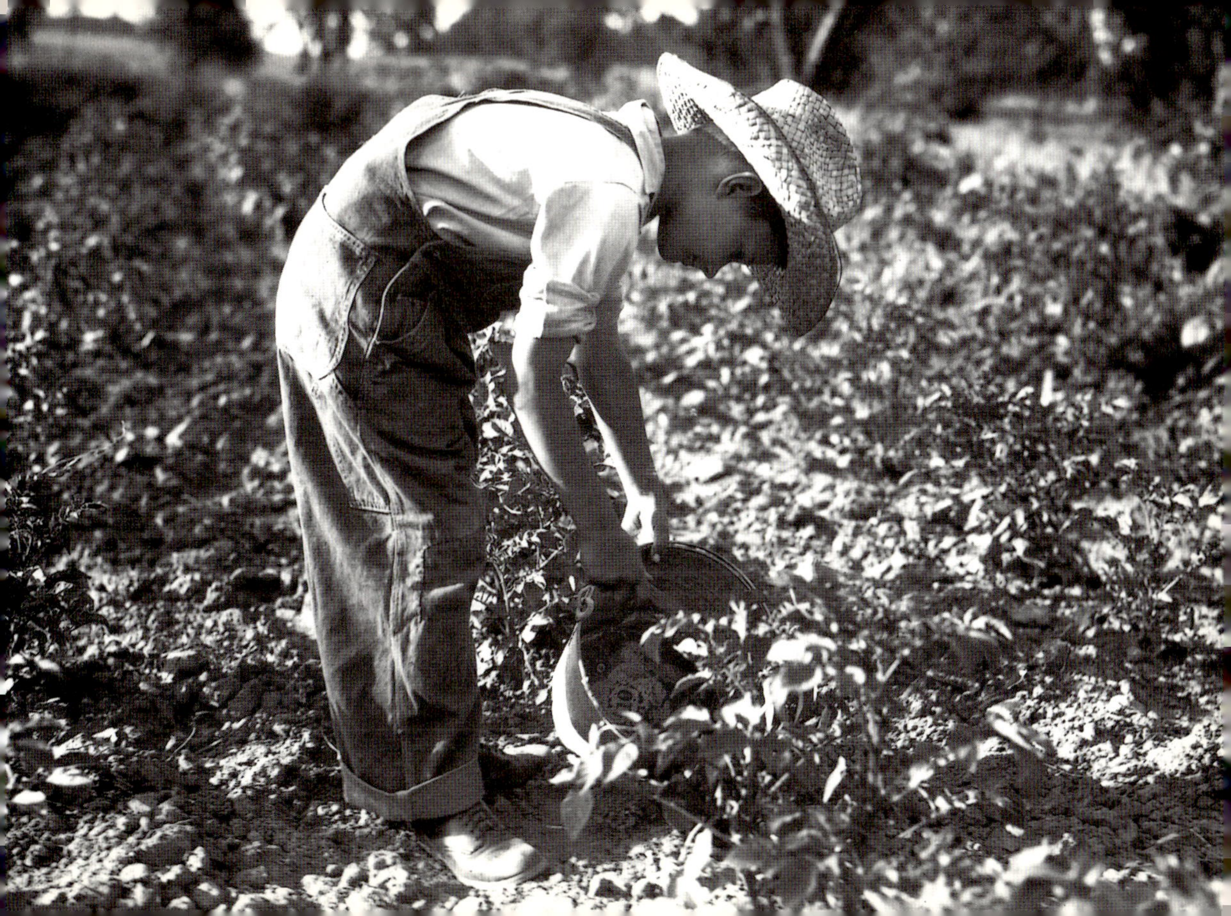

CHILDREN ON THE FARM

"Bugging 'Taters." Knocking potato bugs into a bucket, 1936. Photo by A. M. "Pete" Wettach. State Historical Society of Iowa, Iowa City.

University of Iowa Press http://www.uiowapress.org

USE

FIRST-CLASS

POSTAGE

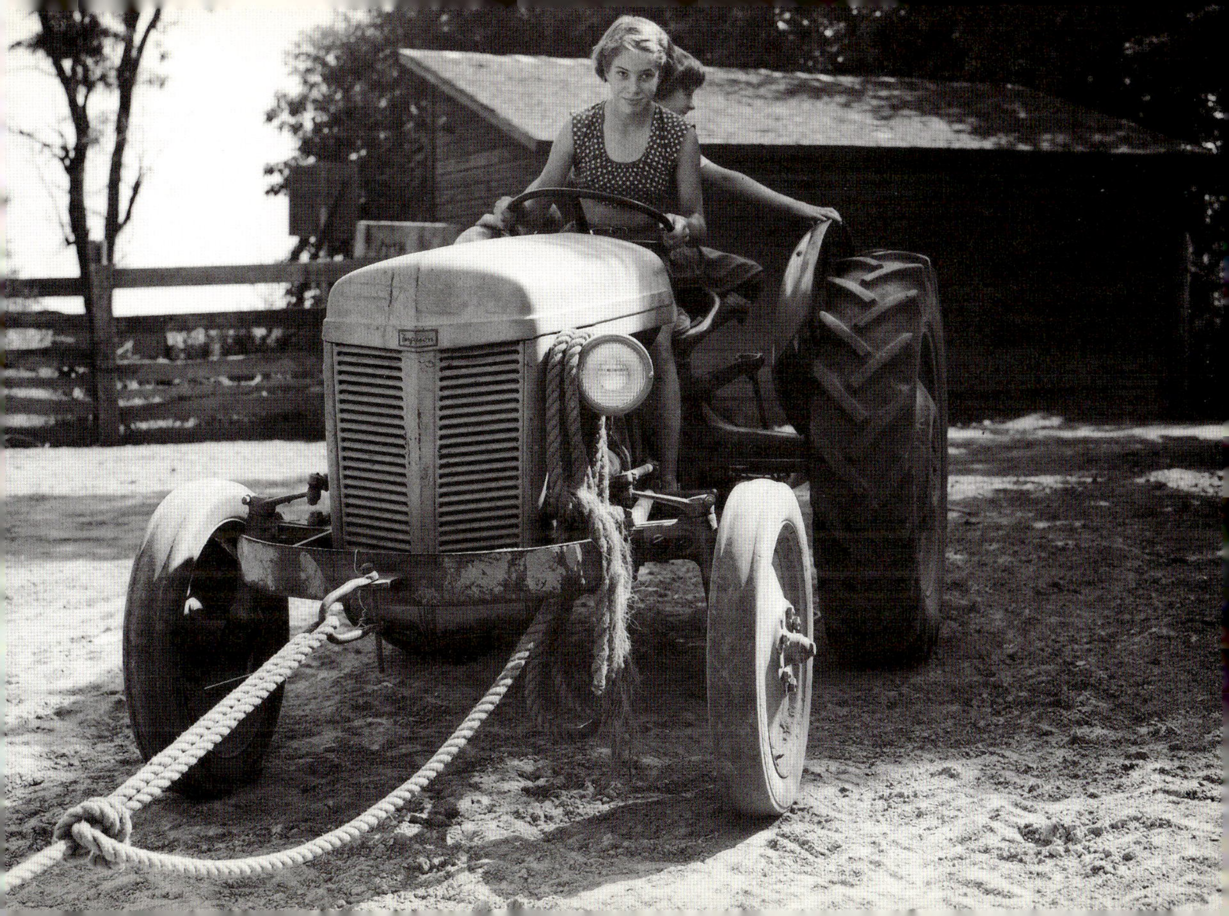

CHILDREN ON THE FARM

Backing up the tractor to pull the hay rope. Photo by A. M. "Pete" Wettach. State Historical Society of Iowa, Iowa City.

USE

FIRST-CLASS

POSTAGE

University of Iowa Press http://www.uiowapress.org

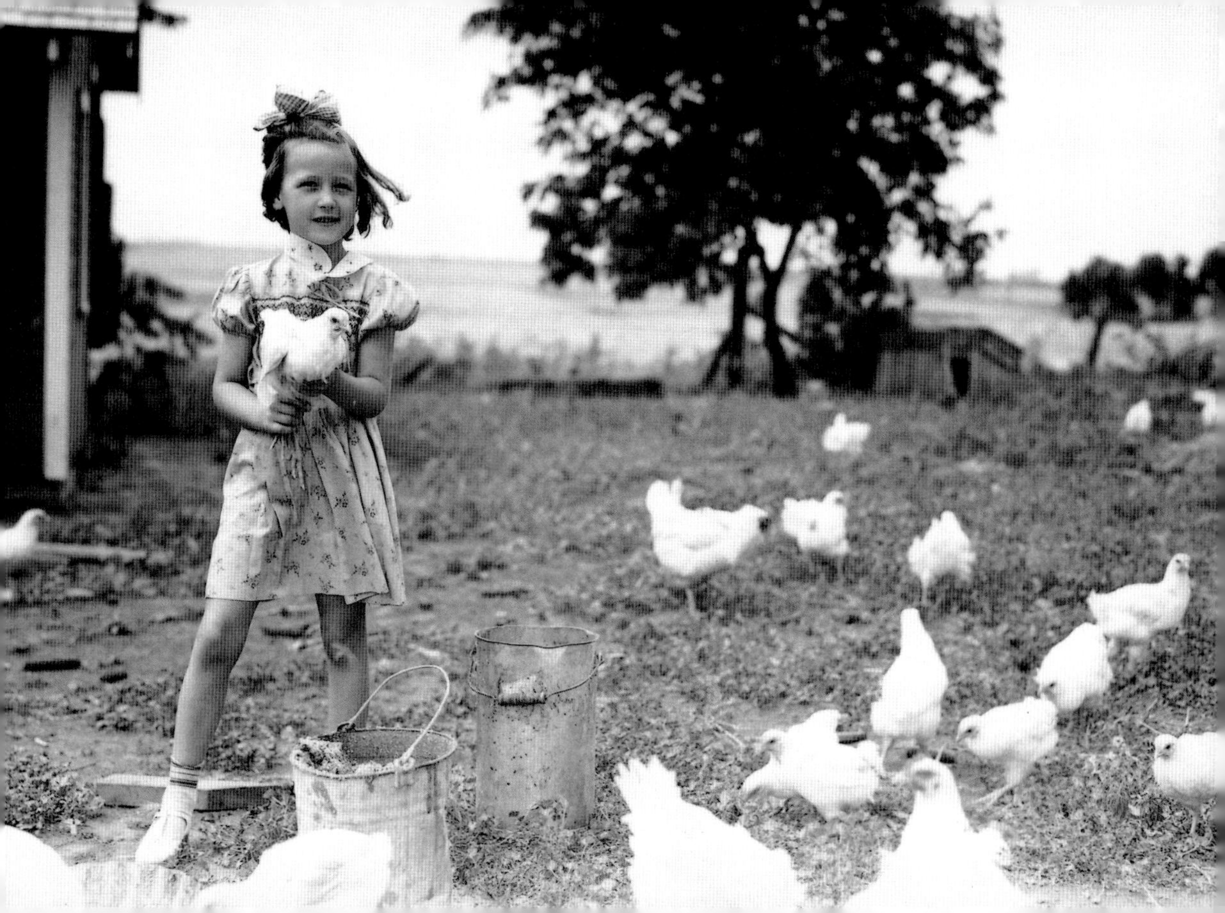

CHILDREN ON THE FARM

Girl holding a chick, Emmet County, Iowa, 1940.
Photo by A. M. "Pete" Wettach. State Historical Society of Iowa, Iowa City.

USE

FIRST-CLASS

POSTAGE

University of Iowa Press http://www.uiowapress.org

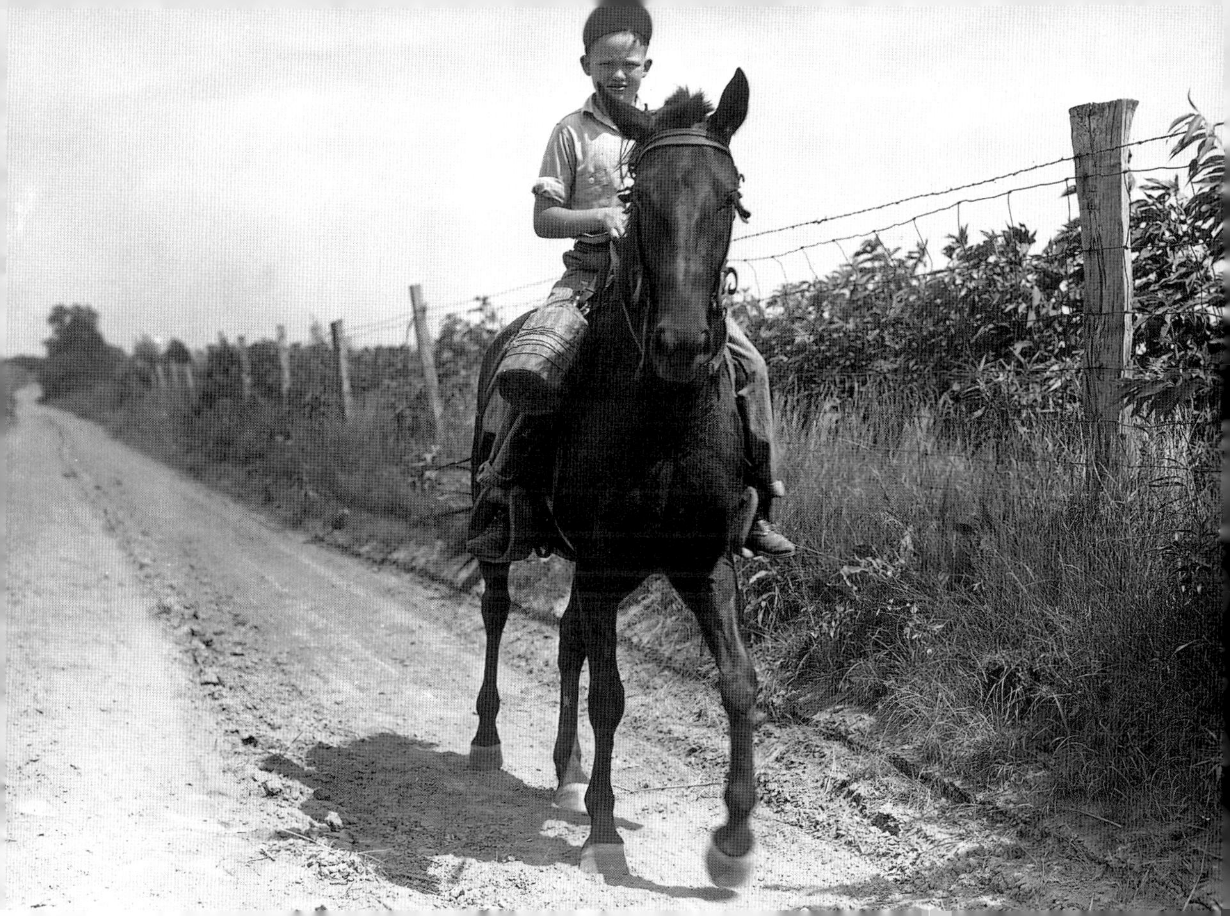

CHILDREN ON THE FARM

Carrying water to field workers, 1939. Photo by A. M. "Pete" Wettach. State Historical Society of Iowa, Iowa City.

USE

FIRST-CLASS

POSTAGE

University of Iowa Press http://www.uiowapress.org

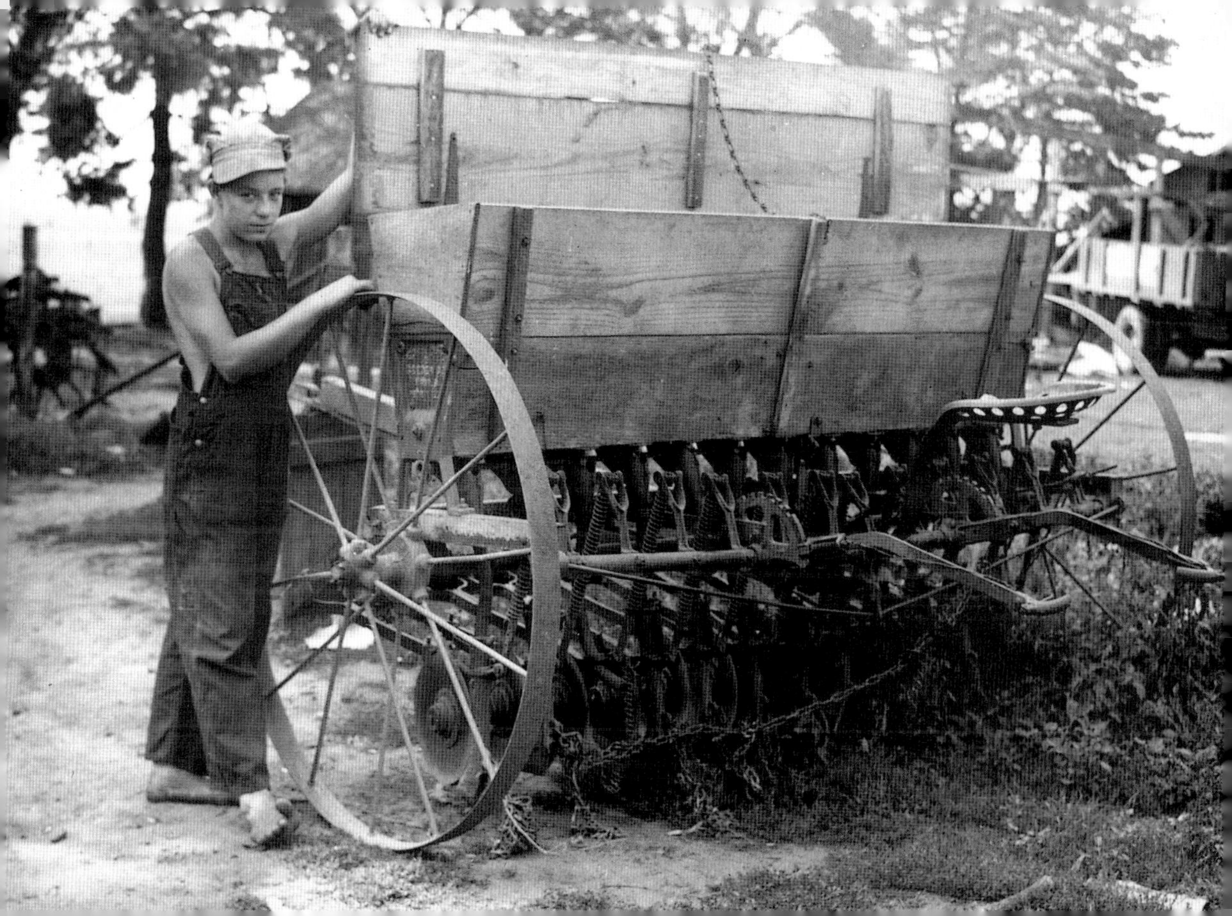

CHILDREN ON THE FARM

A horse-drawn implement converted to a tractor hitch. Photo by A. M. "Pete" Wettach. State Historical Society of Iowa, Iowa City.

USE

FIRST-CLASS

POSTAGE

University of Iowa Press http://www.uiowapress.org

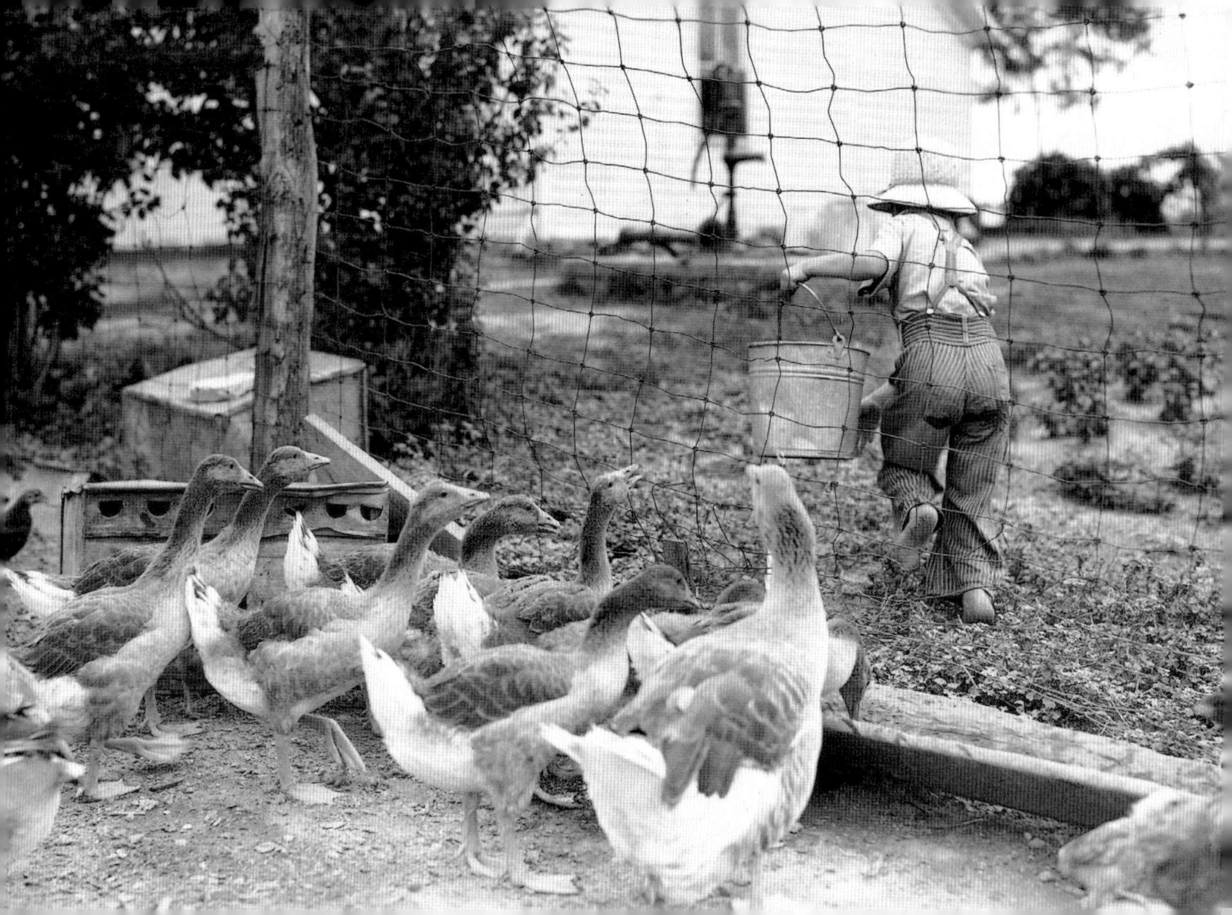

CHILDREN ON THE FARM

Feeding the geese, late 1930s. Photo by A. M. "Pete" Wettach. State Historical Society of Iowa, Iowa City.

USE

FIRST-CLASS

POSTAGE

University of Iowa Press http://www.uiowapress.org

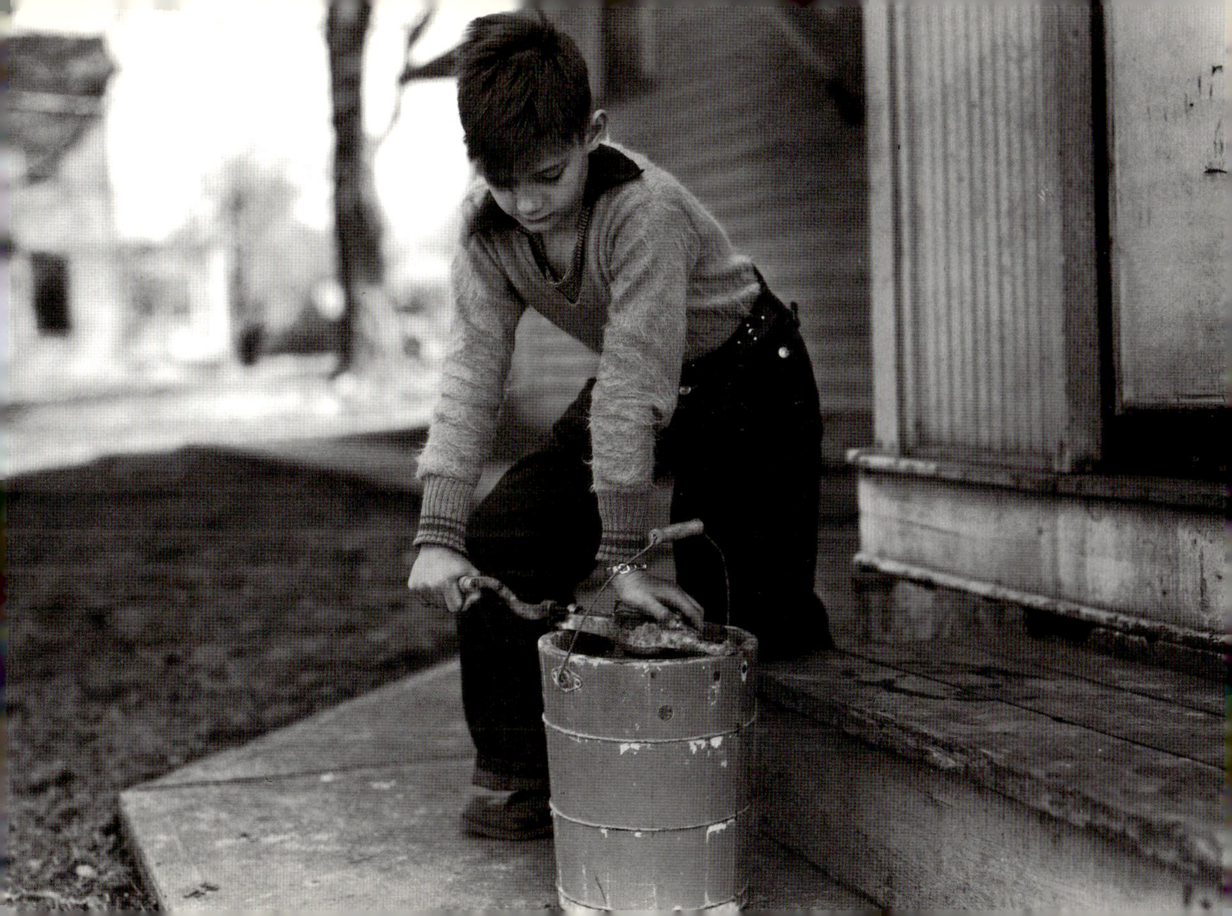

CHILDREN ON THE FARM

Making ice cream, late 1930s. Photo by A. M. "Pete" Wettach. State Historical Society of Iowa, Iowa City.

USE

FIRST-CLASS

POSTAGE

University of Iowa Press http://www.uiowapress.org

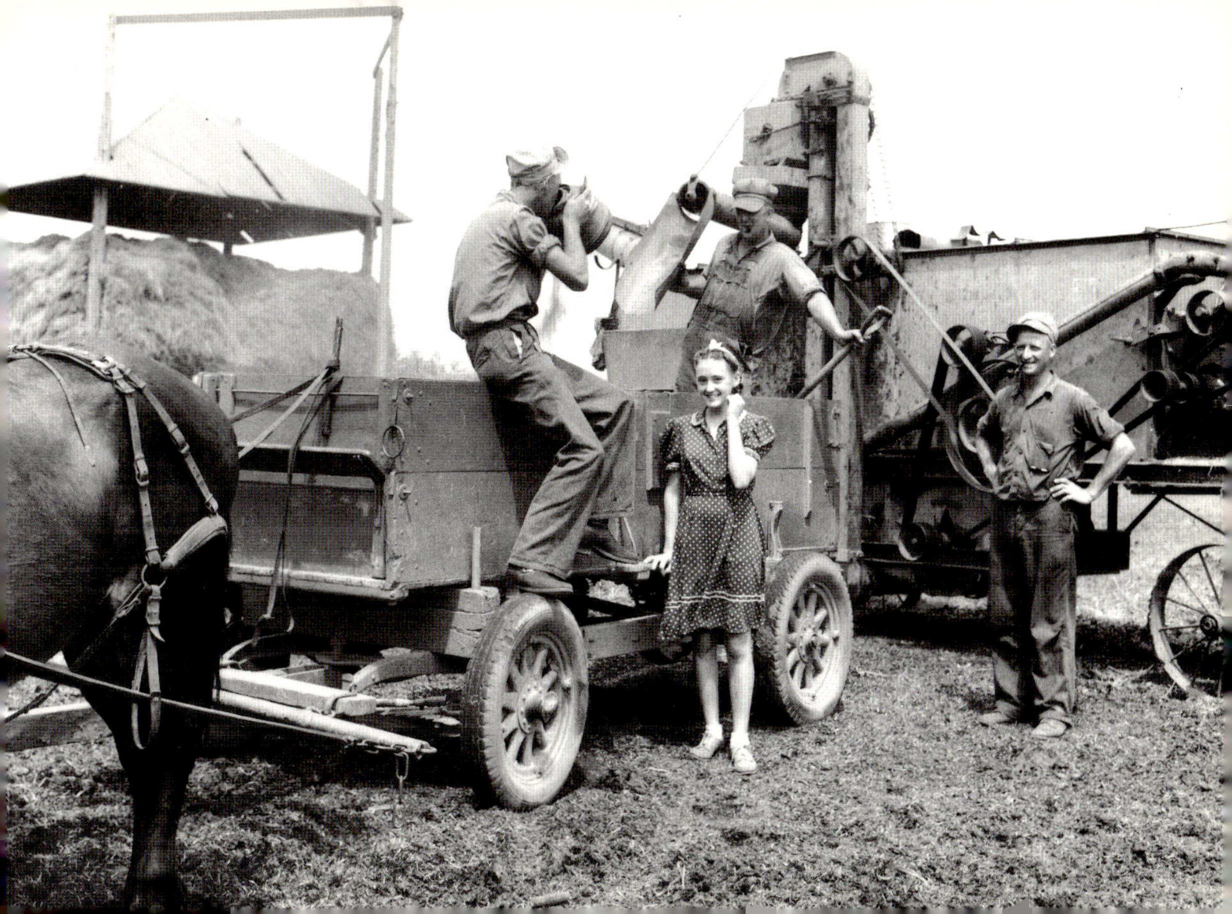

CHILDREN ON THE FARM

Bringing water to the threshers, ca. 1940. Photo by A. M. "Pete" Wettach. State Historical Society of Iowa, Iowa City.

USE

FIRST-CLASS

POSTAGE

University of Iowa Press   http://www.uiowapress.org

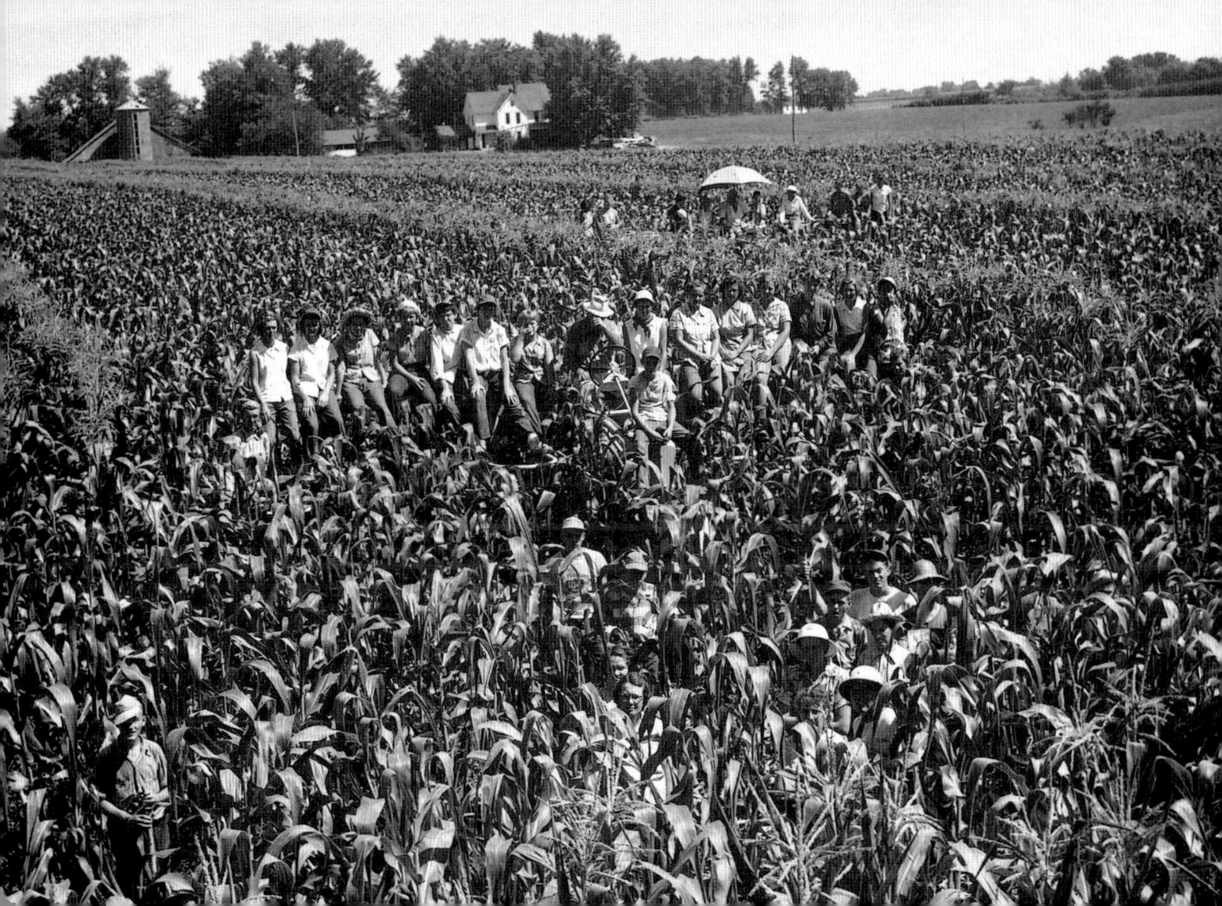

CHILDREN ON THE FARM

"Corn detassellers. School boys & girls," ca. 1960.
Photo by A. M. "Pete" Wettach. State Historical Society of Iowa, Iowa City.

USE

FIRST-CLASS

POSTAGE

University of Iowa Press http://www.uiowapress.org